The Cat in the E Explained

By Jerry Branson

The Cat in the Box Explained

By Jerry Branson

Branson, Jerry
The Cat in the Box / Jerry Branson
Illustrations: Jerry Branson

ISBN-13: 978-0692325964 (Early Learners Publishing)

ISBN-10: 0692325964

1. Children's Books 2. Physics 3. Education

Created and manufactured in the United States.

Acknowledgments

There is a large number of people without whom I could not have written this book. I won the mentor lottery. For some reason, a large number of unusually bright and gifted and talented people have given me their time over the last 50 years to teach me things. I cannot possibly name them all. You know who you are. Thank you *very* much.

Brian Park and Richard Garwin have been instrumental in teaching me concepts specific for this book (and much much more). Frank Dimeo and Brian Rosinski also taught me significant subject matter specifically for this book.

April Rosse, Frank Dimeo, and Lauri Whitehead spent countless hours proofreading and improving this book. Many others have helped proofread sections of the book including Paul Horowitz, Beth Branson, Grant Umeda, Michael Stewart, Mark Crain, Ray Benjamin, Marco Mason, and Mike Curtis.

Special thanks to my father Dick Branson for nurturing everything science and engineering as long as I can remember. Special thanks to my son Sam Branson for writing the wonderful book children's science that got this book started.

Dedication

This book is dedicated to all my mentors who have paid it forward to me with no expectation of return from me and to the children of all ages who learn from this book. Be sure to pay it forward.

License Acknowledgments

Cat model was adapted from Gwinna's *Lowpoly Siamese Cat* model on blendswap.com, under the creative commons attribution 3.0 license.

Dog model was adapted from Priide's *Bulldog* model on blendswap.com, under the creative commons attribution 3.0 license.

Resistor model was used from Someone's *Resistor* model on blendswap.com, under the creative commons attribution 3.0 license.

LED model was used from SonnySee's *Light Emitting Diode – LED* model on blendswap.com, under the creative commons attribution 3.0 license.

PET/CT scan Illustration used with permission courtesy of Roche, Roche.com

Passages from Chad Orzel's *How to Teach Physics to Your Dog* used by permission from Simon and Schuster.

Reprint of Jolly Roger Bradfield's *The Flying Hockey Stick* with permission from Mr. Bradfield and Purple House Press.

Table of Contents

Illustration Index

Table of scientific integrity

Introduction

This book is a companion to the Children's book *The Cat in the Box*, ISBN 0692316205, written by my son, Samuel Branson. His book is a light-hearted rhyming story book version of Schrödinger's Cat aimed at is giving young children some experience in quantum physics to help them understand it when they are older. Recognizing that children ask questions and that not all parents have specific expertise in modern physics, I thought it would be a good idea to offer parents a set of explanations.

It is important to understand that this is not meant to be a physics text book. There are many fine physics books available, and this isn't one of them. The purpose of *The Cat in the Box* is to give children some experience in quantum effects and an opportunity to ask questions about it. The purpose of *The Cat in the Box Explained* is to provide background for parents to answer to these questions and stimulate further interest and questions. We hope that as the curious child matures, she will steal her parent's copy and read the explanations herself.

The concept for the *Mom, I wanna be a Scientist* series started when Sam was in tenth grade and wrote a paper on entanglement, an advanced modern physics concept where whatever happens to one object instantaneously happens to another object at a distant location[1]. We had a through-the-night discussion about how hard this stuff was to understand. It's hard for anyone to understand, even for the people who do this for a living. Even the most eminent physicists do not find it as intuitive as some of us find addition, algebra or even calculus. Rather, they follow the math and find that it gives the correct results, agreeing with experiments and making accurate predictions. Any attempts to link it to our intuition usually fail. J. Haldane said something like, "Quantum physics is not just stranger than we imagine, it's stranger than we can imagine." [2]

Over the next few weeks we got talking about why this stuff is so hard to grasp. One of the reasons, we reasoned, is that when we come across something new, we attempt to apply our everyday experience to it. Thus, our experience since we were infants becomes our guide to understanding unknown things. These experiences become our intuition, and it's difficult to understand things which run counter to our intuition. Quantum physics and relativity definitely run counter to our everyday experience.

---

1 Entanglement is described in the *Busting Quantum Theory* section of this book

2 The actual quote is "I have no doubt that in reality the future will be vastly more surprising than anything I can imagine. Now my own suspicion is that the Universe is not only queerer than we suppose, but queerer than we can suppose." - J. B. S. Haldane, *Possible Worlds and other Places,* 1927.

The things we learn from the time we are young children become obvious to us, whether they are right or they are wrong. Through experience, for example, we learn about traveling.

Suppose Joe is driving north at 25 miles per hour and passes you sitting on a park bench at the center of town at noon. Where will he be an hour later at 1:00? It's "obvious" to most of us that you and Joe will agree that he is 25 miles north of you. If there were a mile marker 25 miles north of you, you would observe him passing this post at 1:00. Joe would agree.

This is obvious because we've had so much experience with this. But, according to Einstein's theory of relativity, this isn't the whole story. If you and Joe measured the distance and times very carefully, you and Joe would not agree on how far he went or how long it took. If Joe had gone very fast, the difference would be large and it turns out that you and Joe would not agree on the distance between you and the mile post or how long it took him to get there. To observe the error, though, Joe would have to travel really fast. Even if he were driving 100 miles per hour, the difference would be so small you and Joe wouldn't notice it. Suppose, however, that he were driving by you at one tenth the speed of light, 18,600 miles per second, and traveled to the mile maker. That's fast enough to go most of the way around the world in one second.

If you measure his trip, you would find him 25 miles north of you about a thousandth of a second later. But if *he* measured the trip, he would claim that he traveled almost a percent less than 25 miles and that it took him only about 99% of the time you measured. If he were traveling at 90% the speed of light, or 167,400 miles per second (which would take you almost to the moon in a second), he'd measure and claim that he traveled less than 11 miles and that it took him less than half the time you measured. But since we don't normally travel so fast in everyday life, we don't experience this length contraction and it is not intuitive or obvious.

Something does not have to be intuitive to be useful, though. Relativity is a good example of a theory which can benefit those who accept it. You may or may not "believe" in relativity, but experiment confirms the math. GPS receivers like the Garmin Nuvi in my car use relativity and can determine where you are to a precision of a foot or so. If they did not take relativity into account, your GPS system will be off by at least hundreds of miles[3]. So you may or may not believe

---

3   For a nice description of the error from a non relativistic GPS system, see *Real-World Relativity: The GPS Navigation System* at http://www.astronomy.ohio-state.edu/~pogge/Ast162/Unit5/gps.html where they show that without taking relativity into account, "the error would accumulate at a rate of about 10km per day. This kind of accumulated error is akin to measuring my location while standing on my front porch in Columbus, Ohio one day, and then making the same measurement a week later and having my GPS receiver tell me I am currently about 5000 meters in the air over Detroit."

relativity, but if and only if you accept it, you can build things that you cannot otherwise build[4]. On this basis, we accept relativity, but it's certainly not obvious or intuitive to most of us.

Let's take this one step further. Heisenberg's uncertainty principle says that the more you know about how fast you are going, the less you can know about where you are. If you know exactly how fast you are going, you can know nothing about where you are. This goes beyond not being intuitive – it is actually *counter-intuitive* because it is in direct conflict with our experience. We have all stopped at a red light, knowing exactly where we were and that we were exactly at a dead stop. When Heisenberg's principle claims that we can't know exactly where we are and exactly how fast we are going, we know better! It is for this reason that we can't truly understand some of these things at all.

Let's look at an example that became obvious over time: does something heavy always fall faster than something light? Until 1586 most people thought that it did. Their experience with heavy things (like rocks) was that they fell fast and their experience with light things (like feathers) was that they fell slowly. It was obvious to everyone that heavy objects fell faster than light objects, as described nearly 2500 years ago by Aristotle (who's going to question Aristotle?!?).

Nevertheless, it is obvious to nearly everyone these days that if there's no air resistance heavy objects fall at the same rate as light objects. Why the change? Because we were told of Galileo's experiment at a very early age. We may have also read about, or seen live on TV when Neil Armstrong repeated the experiment on the moon 2500 years after Galileo. In 1969, it was considered more "cute" than surprising. When Sam was very young, he and I actually dropped a feather and a marble in a vacuum, a six foot long Plexiglas tube with the air removed. A feather and rock falling at the same rate became natural and obvious to him. He learned early on about air resistance.

When I was young, one my favorite authors was Jolly Roger Bradfield. One of Bradfield's books is called *The Flying Hockey Stick*. My mom must have read this book to me 1000 times. After years of engineering school and decades of engineering experience, I have a reasonably good handle on

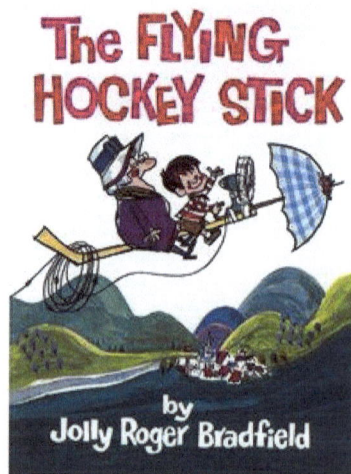

Illustration 1: Learning physics from a childhood favorite

classical physics. I understand and routinely apply Newton's laws of motion. Still, though, I have to fight the intuition of Bradfield's picture, because it seems obvious

---

4   But this is not the whole story. See the section *Why is this egregious error still in the book?*

that the hockey stick will be propelled forward. Why? Because I was exposed to Bradfield physics from an early age. What your mom tells you becomes intuitive.

In *How to Teach Physics to your Dog*, Chaz Orzel says, "Dogs come to quantum physics in a better position that most humans. They approach the world with fewer preconceptions than humans, and always expect the unexpected. ... If dog treats appeared out of empty space in the middle of the kitchen, a human would freak out, but a dog would take it in stride. ... Quantum mechanics seems baffling and troubling to humans because it confounds our commonsense expectations about how the world works."

One night Sam and I asked ourselves, "What if someone had been exposed to and given experience in quantum physics from an early age?" They may very well be able to comprehend quantum physics from an intuitive viewpoint and solve problems that no one is currently able to solve. Our aim with the *Mom, I want to be a Scientist* series is to give young people experience and expectations which are in line with modern physics. We don't need to lie to them about the macro world, but just give them stories about traveling into the microscopic world. In this way we hope to prepare some humans to be even more equipped to master quantum mechanics than dogs.

We're not proposing to teach a preschooler quantum physics and relativity. Rather, we want to expose preschoolers to their consequences. If some of your childhood books talked of Joe returning from his trip to Proxima Centauri and being younger than his twin brother, the concept of time dilation would not be so unbelievable during a college course on modern physics.

So this is our aim – to raise a generation of people well prepared to take on modern physics. This cannot be done in a single reading of a single book. Rather, it will take years of development through increasingly deep exposure to a series of books ranging from Dr. Seuss style to Harry Potter level where we develop intuition in the same way I developed an understanding of Bradfield physics. How many of us at one time thought if we fell off a cliff we would start falling only after we looked down and realized we should be falling? Thank you Wile E.

This book, *The Cat in the Box Explained*, is a companion to *The Cat in the Box*, the first in the series. Each page teaches several concepts to be savored over a number of years. On the surface, it is a children's story of Schrödinger and his friend Einstein, trying to figure out what kind of pet he has.

This book is divided into categories, designated with green, yellow, blue, and red bars on the side of the page. There is a single green page for each scene in the book with light reading background discussion material. The yellow sections go deeper into the subject, often with some light math. The blue and red sections contain advanced information.

One way to read this book is breadth first – read all the green pages first, skipping the other pages. When you want more detail on a particular section, you can proceed to the yellow, blue, and red pages.

In some professions, a school's main goal is to teach you how to do your job. Accounting, dentistry, and air conditioner repair are examples of theses – of course you learn more as you go and get better at it, but the school's goal is to have you ready to be productive immediately out of school.

This is not true for all professions, though. In many branches of engineering, for example, the goal of the school is to teach you how to think in such a way that you are prepared for learning efficiently. You will learn what you need to know to do a specific engineering task when the times comes. It is not practical to teach you everything you need to know to do all possible engineering tasks – you could go to school for 40 years and not be able to do this.

Loosely speaking, that is the philosophy of this book. I give examples to keep the reader's attention and keep things interesting, but this is in no way a comprehensive course. Rather, I am teaching how to think in a certain way to prepare someone for learning Science Technology Engineering and Math concepts.

So this our goal – to give young people experience to understand currently un-understandable concepts. By giving people the correct experience from an early age, concepts like quantum physics and relativity need not be counter-intuitive.

Most of the quotes on the back cover are tongue in cheek, but not all of them. Richard Garwin's comments are real and speak well to my overall goals. I do not pretend to understand this subject (few people lay claim to this). I take seriously Pief's words which apply very well to this subject, "What we don't understand we explain to one another." This book obviously does not cover every possible question on modern physics. Additionally, this book will no doubt stir some discussion. In this vein, we've set up a discussion web site to have discussions on Sam's book, these explanations, and the subject in general. A link to the site can be found on the *Mom, I wanna be a Scientist* series website, www.earlylearnerpublishing.com

An additional fun aspect is that the series logo described in the *Light Cones and Wavelets* section of this book is hidden on each page of Sam's book. I'll reveal the secret locations in the green sections of this book.

The importance of Diagrams

The paper the dad is giving his daughter in Illustration 2 is called a Feynman Diagram. A Feynman Diagram is a picture version of the complex math behind particle physics. Richard Feynman invented them in 1948 to make it easier to find the probability that something would happen at the quantum level. He won a Nobel prize for this work in 1965.

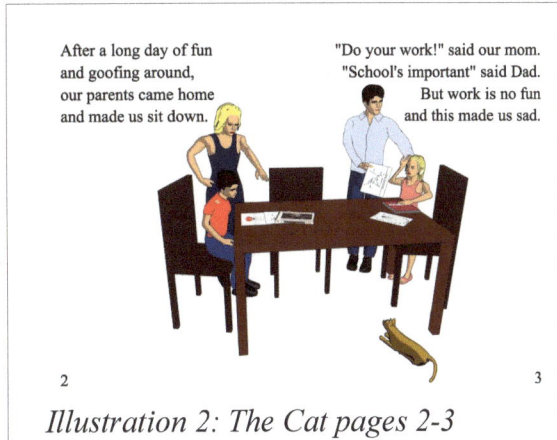

Illustration 2: The Cat pages 2-3

There were two major contributions Feynman made when inventing his diagrams: 1. He proposed an interpretation of the positron as if it were an electron moving backward in time. 2. He made a relatively simple diagram that replaced some very messy and difficult to handle calculations allowing people to visualize the problem and solution rather than relying solely on a vigorous abstract mathematical representation.

This is the type of genius that allows understanding to move forward when a subject has reached complexity equal to the limit of human understanding. This is the type of understanding, cleverness, and insight this series strives to create.

We'll start with an example of how to calculate something for a video game and see how using visualization tools such as the graphing you learn in school can make understanding much easier. Feynman was a master of such tools. We will return to this example of diagrams and alternate interpretations of time at the end of this book and see how this seemingly simplistic technique can be used to understand advanced concepts.

Feynman was a character. In addition to being a truly extraordinary genius, (see *No ordinary genius* by Christopher Sykes, ISBN 039331393) he was an adventurous guy and wrote a lovely autobiographical book *Surely You're Joking, Mr. Feynman!* (ISBN 978-0393316049) which brought him to fame and shows the excitement and joy he brought to himself and physics. This light book is infectious and makes you want to become a physicist. *It is a must read.*

The series logo is on the paper Tim is reading.

Where's the explosion?

Warning: This first section contains math. Please don't let it turn you off. There is very little math in this book, but I need to use it here how to demonstrate how to understand concepts when the math can seem daunting. Please bear with me for just a few pages.

Consider a problem you encounter when writing a video game. The player has fired a laser toward a wall and you need to know when and where on the screen to draw the explosion. So, basically, you want to find out where the wall and the laser beam intersect. You can do it by using algebra and isn't too difficult once you fully understand the problem and its mathematics. It is then a matter of solving the algebra and writing the software.

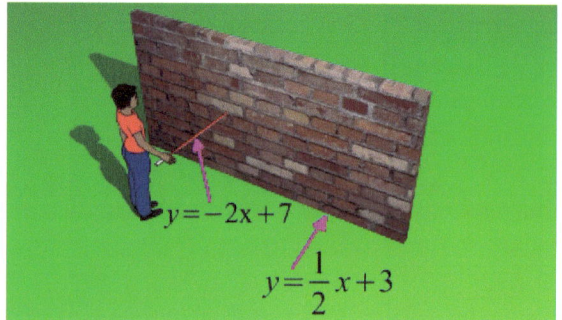

Illustration 3: Shooting a laser at a wall

$$y = -2x + 7$$

$$y = \frac{1}{2}x + 3$$

But if this is the first time you're doing something like this, the mathematics can be daunting. To understand the problem, some drawings and a graphical solution provide an excellent way to learn about the problem and develop the solution. We will show this over the next few pages.

First, take a minute to jot down how you might go about determining where you should draw the explosion. You don't have to solve it, but just write down some ideas about how you might go about attacking this problem. If nothing else, jot down what goes through your mind when you think about the problem for a few minutes.

_____

_____

_____

_____

_____

_____

Suppose you know the equation for the location of the wall as

$$y = \frac{1}{2}x + 3$$ inches and the path

of the laser as $y = -2x + 7$ inches and that Tim's hand is at the point (x=3½, y=0). Let's assume that Tim's hand is at the same height as the spot on the wall where the laser hits. That way, we can solve for x and y without worrying abouth height z.

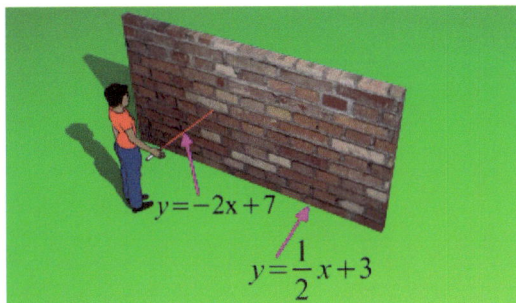

Illustration 4: Equations of the laser and wall

Illustration 5: Representation on a graph

This can be thought of as taking the picture of the wall and superimposing it on some graph paper, and putting the paper at the same height as Tim's hand as shown in Illustration 5:

Suppose the laser beam moves toward the wall at speed of ten inches per second, like in movies. You need to simultaneously solve the two lines and draw an explosion at this location on the screen just as the laser beam hits the wall. You can solve these equations in a variety of ways.

Using Algebra to find the explosion:

You could, for example, take the following algebraic steps:

1. The goal is to find the location (a value for X, Y), and time T which satisfy the equations. Let's find y first. To do this, we will eliminate the x terms.

Multiply the wall equation by 4 so the x term is the same size as in the other equation:

$$y = \frac{1}{2}x + 3 \quad \rightarrow \quad 4y = 2x + 12$$

2. Add the two equations together, eliminating the x term:

$$\begin{aligned} y &= -2x + 7 \\ 4y &= 2x + 12 \\ \hline 5y &= 19 \end{aligned}$$

3. Divide both sides by 5 to solve for y:

$$y = \frac{19}{5} = 3.8$$

4. Substitute 3.8 for y into one of the equations to find x:

$$3.8 = \frac{1}{2}x + 3$$

5. Subtract 3 from both sides to eliminate the constant on the side with x:

$$0.8 = \frac{1}{2}x$$

6. Multiply both sides by 2 to find x:

$$1.6 = x$$

7. Check by putting the solution into one of the equations to be sure we didn't make a mistake:

$$\begin{aligned} 3.8 &= (-2)(1.6) + 7 \\ 3.8 &= -3.2 + 7 \\ 3.8 &= 3.8 \end{aligned}$$

We can now find T. Using the distance formula to find out how far Tim's hand is from the wall,

$$d = \sqrt{(x_2 - x_1)^2 + (y_2 - y_1)^2} = \sqrt{(3.5 - 1.6)^2 + (0 - 3.8)^2} = \sqrt{3.61 + 14.44^2} = 4.24$$

Since the laser travels at ten inches per second, it would take 0.424 seconds to go 4.2 inches. We now know the explosion should be drawn at (1.6, 3.8) 0.424 seconds after the laser is fired.

So now you know to draw the explosion at (x=1.6, y=3.8) 0.424 seconds after the laser is fired. This may seem like a lot of work.

Using a graph to find the explosion:

Does the algebra look like a lot of work? Could this be confusing to someone who hasn't done it before? You betcha! Here's an alternate method: draw (plot) both objects (lines) on the same graph and simply *look* at where they intersect as in Illustration 7.

Illustration 6: Converting a drawing to a graph

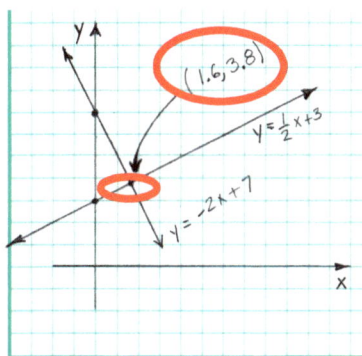

Illustration 7: Finding a solution with a graph

If you don't know how to solve the problem, determining how to do it algebraically is a pretty tall order and is the sort of thing that might make someone hate algebra, or even hate math. But if someone showed you the graphical solution first, you may find the algebra exciting, especially if you really wanted that video game.

A graphical solution is approximate – sometimes good enough for what you're doing, great for quickly checking your other solution, and always good for learning and determining how to find more exact solutions. That's the reason your math teachers want you to draw a picture with each problem.

This video game example is kind of what Feynman diagrams did for quantum electrodynamics, a branch of quantum physics dealing with light. Just as the graphical solution is easier to both visualize and find the location of the explosion, following the Feynman diagrams makes it a *whole* lot easier to visualize and solve than the mathematics which require the use of large and complicated integrals over a large number of variables. Feynman took concepts that were highly abstract and found a way to see them graphically to understand and solve them without needing such a high level of mathematical abstraction.

Feynman invented his diagrams when almost *no one* understood quantum mechanics. Feynman diagrams made quantum physics accessible to a whole lot more people.

Atoms

Suppose you have an apple and you cut it in half. You take one of those halves and cut it in half to get a quarter of an apple. Then you cut the quarter apple in half and get an eighth of an apple.

If you continue cutting the apple, after about 20 cuts you will need a magnifying glass to see it well enough to cut it. After a few more cuts, you'll need a microscope. But you can still cut it. How long can you continue cutting in half before the piece can no longer be cut in half? Can you do it forever?

This is a question that was contemplated for a very long time. Some people imagined the apple to be made of something like jello that could be divided forever while others imagined it as if it was made up of tiny Lego pieces and could only be taken apart so far before you came to indivisible, or elementary pieces.

It turns out that everything is made out of building blocks and there is a limit before you have elementary pieces that you can no longer cut. It was Albert Einstein who provided the proof that this was the case. We will explore this in more detail in the section "The miracle year of 1905."

How many cuts will be made before you get to a single, uncuttable piece? The answer for an apple is somewhere around 82 cuts before you get one atom, a fundamental building block which can no longer be cut in half. It's surprisingly few cuts! For the math savvy, $\log_2(6.022 \mathrm{x} 10^{23}) \approx 79$ and as a check, $2^{79} = 6.045 \, x \, 10^{23}$. What the check means is that if I take one atom, double it and get two, double that and get four, and after performing 79 doublings I will have $6.045 \, x \, 10^{23}$ atoms. If they were all carbon, that would weigh about 12 grams. A small apple weighs about 100 grams, so we need to double the 12 grams about three more times. Hence the number 82. A very large apple is only about double a small apple, so it might take 83 cuts. Still *much* lower than I was expecting before I calculated it.

For a while, it was thought that they had found the elementary particles and called them atoms. Two examples of atoms are hydrogen and oxygen. You can make water by combining hydrogen and oxygen.

Eventually, someone determined that atoms were not really elementary – they were made of smaller building blocks: protons, neutrons, and electrons. The center of the atom is called the nucleus and is made up of protons and neutrons. Surrounding the nucleus are electrons. Protons and neutrons are about the same size and weight while electrons are much smaller and weigh only about one thousandth as much as a proton.

The chemical characteristics of atoms – how they react in chemical reactions – is determined mostly by the number of protons in the atom, so we characterize atoms by the number of protons. There are 83 different types of atoms found naturally on earth. We have both named and numbered the atoms.

The simplest atom is hydrogen with one proton, one electron, and no neutrons. We often designate the number of protons, or the atomic number, as z. Therefore, $z=1$ for hydrogen atoms. We also abbreviate the name of the chemical with one or two letters. In the case of hydrogen, we use a capital H.

An atom

Atoms are often drawn as a tiny solar system with the electrons orbiting the protons and neutrons. This overly simplistic model is not suitable for modern physics. For the time being, I'd like you to think of the protons and neutrons as sitting in the middle of the atom with a spherical cloud surrounding them. The electrons are somewhere in this cloud. There is a probability associated with finding an electron at any particular place in the cloud. This space is often called the electron probability cloud. We'll talk about this in more detail later in the *Wave functions and probability functions* section.

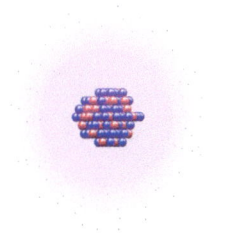

Illustration 8: A common model of the atom

If we add a neutron to hydrogen so we have one proton, one neutron, and one electron, it is still hydrogen and $z=1$. But the weight approximately doubles since a proton and neutron weigh about the same and the electron doesn't weight very much. When we have two atoms with the same same number of protons but different numbers of neutrons, we call them isotopes. To designate the isotopes of an atom we often precede the chemical symbol with its weight. So hydrogen with no neutrons is ^{1}H and hydrogen with one neutron is ^{2}H.

This is the most common form for writing an element – the standard one or two letter abbreviation preceded by a superscripted mass. For hydrogen, that's not too bad. You can probably remember that H stands for hydrogen and that the atomic number for hydrogen is 1. But until you have the periodic table memorized from so much use, this can be a challenge for a lot of elements. For example, the chemical symbol for lead is Pb. Pb is an abbreviation of plumbum, which is Latin for "soft metal." It is atomic number $z=82$. For example, when it has 124 neutrons it is called lead 206 and is designated ^{206}Pb. It's asking a lot for someone new to the subject to remember that ^{206}Pb is lead with 82 protons and 124 neutrons, so I will use a system which includes the number of protons as a subscript, $^{206}_{82}$Pb. This system isn't used very often in books because the people who write the books have the periodic table memorized and they most often use ^{206}Pb.

Everything on earth is made of atoms. There are 83 different types of atoms found naturally on earth and another 30 or so that we have made. Atoms are made out of neutrons, protons, and electrons. When neutrons or protons are bombarded at high speed with other neutrons and protons, they can be broken down further into smaller components called quarks. As far as we know, electrons are fundamental and cannot be broken apart. There are many other particles that have been discovered, so many that when physics giant Enrico Fermi was corrected on the name of a particle, his response was "Young man, if I could remember the names of all those particles I would have been a botanist!" Some other well known particles include photons, muons, gluons, bosons, and many different types of neutrinos. The Wikipedia article *List of Particles* lists enough particles to understand Fermi's response.

Likes and dislikes

There are two types of electric charges. Ben Franklin named them positive and negative. Franklin found that like charges repel each other and unlike charges attract each other. What this means is that two positive charges repel each other – if you try to push them together they fight back and try to prevent you from pushing them together. The same thing happens if you try to push two negative charges together. But if you have a positive charge and a negative charge, something different happens: they attract each other and if you try to pull them apart they fight back and try to prevent you from pulling them part.

Two electrons repel each other

Two protons repel each other

An electron and a proton attract each other

A proton and an electron attract each other

Illustration 9: Attraction and repulsion

8

A magnet has two poles and similar behavior. Two north poles repel each other and two south poles repel each other, but a north pole and south pole attract each other.

Go get two magnets and try this!

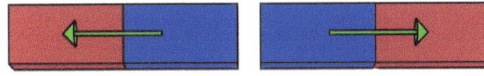

Two south poles repel each other

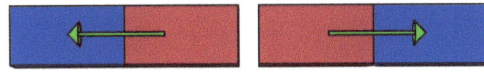

Two north poles repel each other

A south pole attracts a north pole

A north pole attracts a south pole

Illustration 10: Magnetic Forces

If you try this with refrigerator magnets, though, you will get a seemingly different behavior. This is because a refrigerator magnet is not a single magnet with a single north and single south pole. Refrigerator magnets are made up of strips of magnets. If you put them together they will "snap" together such that one's south poles are near the other one's north poles as shown below. If you try to slide the magnets to an arrangement where one's south poles are near the other one's south poles, they will fight you and snap to an arrangement where one's norths are lined up with the other's souths. You should try this!

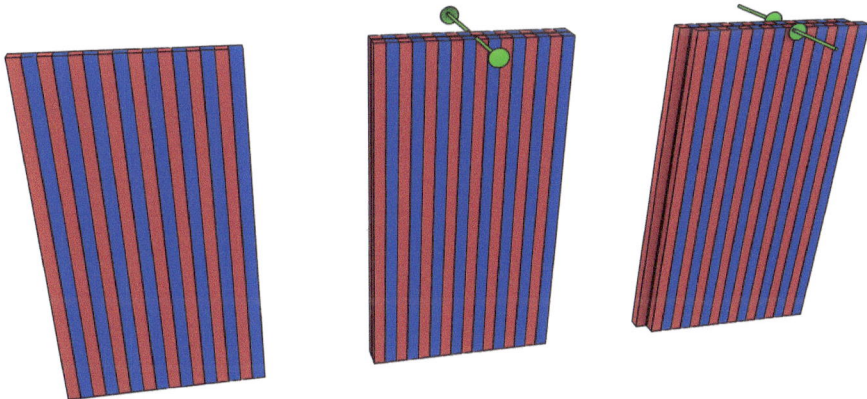

A refridgerator magnet is a bunch of magnets put together

If aligned north to north, two refridgerator magnets repel each other

If aligned north to south, two refridgerator magnets attract each other

Illustration 11: Refrigerator magnets

Seeing the invisible

Illustration 13: My collection of magnets

Keep in mind that magnetic poles are invisible and you won't be able to see the magnetic poles like they are shown in red and blue in Illustration 11. But there *is* a way to see them. You can buy magnetic viewing paper. Just search for "magnetic viewing paper" or "magnetic viewing film" on the Internet. You can buy some for around $10.

Illustration 12: Seeing magnetic fields

To the left you see my collection of magnets on the front of my refrigerator. To the right, I've taped a piece of piece of magnetic viewing paper over the magnets. Most of them look like you might expect, with a field shaped like the outside of the magnet. But take a close look at the hard drive magnet at the top. There are actually two different magnets inside that package. The rectangular refrigerator magnet at the bottom clearly shows the strip pattern.

So how does this work? Inside the paper are some shiny magnetic flakes (they usually use nickel) suspended in dark oil.

When you study the fascinating field of magnetics (pun intended!), you'll start drawing magnetic fields with curves. The magnetic flakes line themselves up in the direction of the field. Where the field is perpendicular to the viewing paper, the flakes stand up on edge and are hard to see so that region appears dark. Where the field is parallel to the viewing paper, the flakes are parallel to the viewing paper and are easy to see so that region appears light.

Illustration 14: Magnetic field of a refrigerator magnet

Years ago I laminated a piece of the magnetic viewing paper and cut it the size of a credit card. I carry it in my wallet. Here's a picture of it being used to show the magnetic field of a refrigerator magnet. This little gem has come in handy many times when I've been somewhere and wanted to know what a particular magnetic field looks like.

For example, here is my wallet viewer showing the field from a magnet I removed from a motor. Note that there is no way to tell by looking at the magnet that the field has this pattern.

Below I've put two of the motor magnets near each other. Look at the pattern between the motors. While I could have calculated what the field looks like, putting a piece of viewing paper over it was *way* easier!

Illustration 15: A speaker's magnetic field

Illustration 16: Added magnetic field of two speaker magnets

Anti-matter

For each particle, there is something called an anti particle that has the same characteristics except opposite charge. So an anti proton has the same mass as a proton, but it is negatively charged while the proton is positively charged. An anti electron (also called a positron) has the same mass as an electron, but it is positively charged while the electron is negatively charged.

When a particle and its anti particle interact, they are both annihilated (destroyed) and their energy is converted to another form, such as a pair of photons.

A *whole* lot of energy is given off, so this would be a really nice way to store energy if you had a supply of anti-matter. This is why in science fiction when writers who don't have to worry about where to get things use matter/anti-matter reactors for energy sources. This would give you billions of billions times the energy density of gasoline which is already about 50 times the energy density of cell phone batteries. If you could use this in a cell phone, you'd buy it and not have to recharge it your entire life.

So why don't we use this technology for cell phones? The energy storage device must be practical for the application and the overhead plant required to use anti-matter is not practical. Consider the fuel tank – what could you make the container out of that would hold anti-matter? The answer is that there isn't one! Any container you made would immediately be annihilated by the anti-matter. So you have to hold it in something called a magic bottle. It's a container made out of magnetic fields that hold the matter without touching it. This container is fittingly called a magic bottle. It's a fascinating subject – do a search a read up on magic bottles.

Another consideration is cost. With today's technology, anti-matter is very expensive to make. How expensive? In 1999, NASA gave a figure of $62.5 trillion per gram of anti-hydrogen. Your Dad might not like that phone bill.

What is light?

Light is a form of energy which our sense of sight can detect. It is made of electromagnetic radiation and travels in a straight path. Electromagnetic radiation is radiation that varies its electrical and magnetic strength as it travels.

This kind of radiation is not normally harmful unless you get too much of it for example, if you look at the sun or if you stay out too long and get a sunburn.

We looked at our sheets.
It was physics alright.
It was one simple question:
What exactly is light?

We were so bored
that we argued for fun!
But Mom wasn't mad
that we got nothing done.

4

5

Illustration 17: The Cat pages 4-5

If you think of light as a wave, the height of the wave indicates how bright the light is and the wavelength (the distance between two crests) indicates the color of the light.

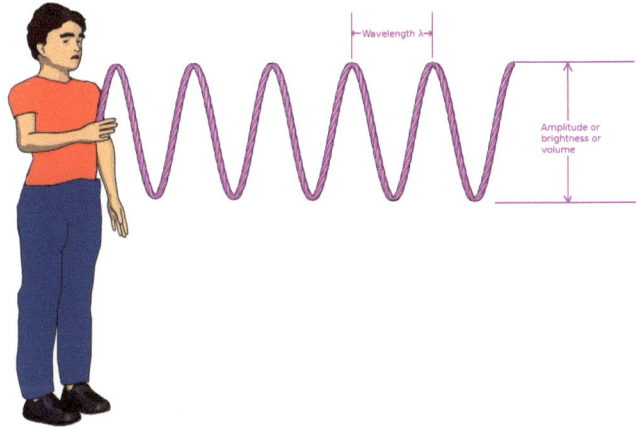

Wavelength λ

Amplitude or brightness or volume

Illustration 18: Amplitude and wavelength of a wave

If you look at more than one wavelength at the same time, you will perceive a combination color. For example, if you look at red light and green light together you will see yellow. If you look at a lot of wavelengths (colors) together, you will see white. Why this happens is explained in our upcoming book about light.

Light travels fast. *Really* fast. Light travels 186,000 miles in one second. In one second it can make it around the earth seven times. Light can make it all the way from Virginia to California and back 40 times in one second. It would take me about six months to do that in my car if I didn't stop even for gasoline. It takes just over one second for light to travel from the moon to the earth. Light travels very fast.

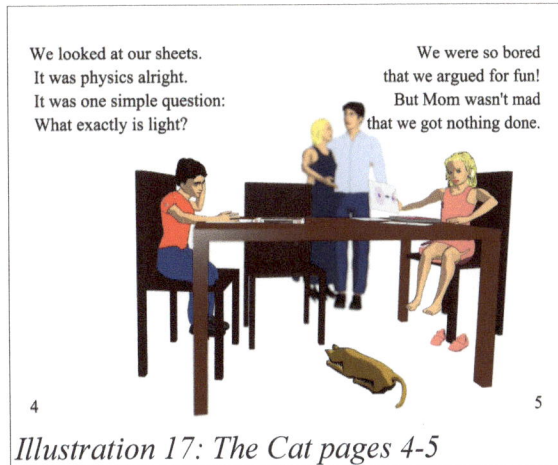 The series logo is carved into the chairs.

13

The energy of photons

All photons travel at the same speed, the speed of light, so when a photon is ejected the amount of energy is not reflected in its speed. Instead, it is reflected in the wavelength of the photon. A relatively low energy photon has a relatively long wavelength – red photons have a longer wavelength than violet photons, so a violet photon is more energetic than a red photon. One way to visualize this is shown in Illustration 19. To get the shorter wavelength, Tim has to move his rope up and down faster than Sally. He is going to get tired faster than Sally because she doesn't have to move her hand up and down as fast.

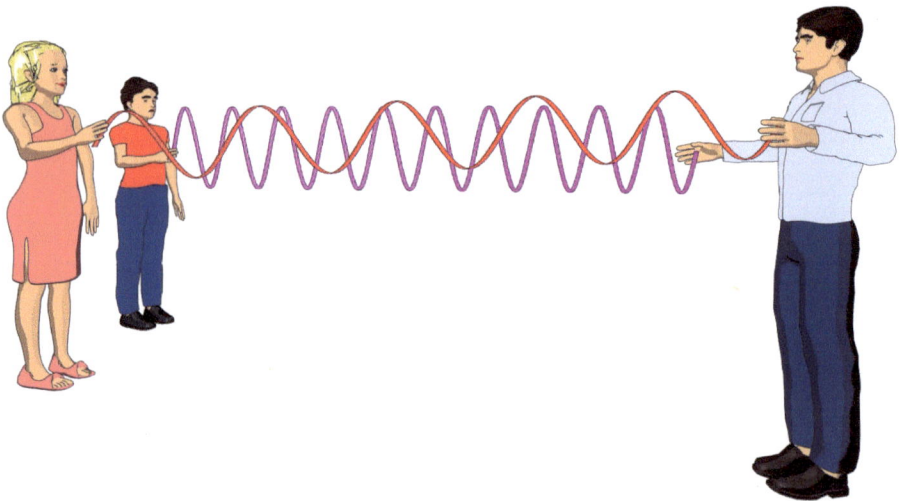

Illustration 19: Purple photons have more energy that red photons

Ultra Violet (UV) photons have even more energy than violet photons, and this is why you need to protection from UV light. Gamma radiation photons usually have a wavelength much much shorter than even UV photons, making them highly energetic, and very difficult to stop and very dangerous to living organisms.

After reading this section, one of my friends asked, "If gamma radiation is just photons of light, why can't I block it with a piece of black tape?"

If the photons are visible, you can. But if the wavelength is short, the "light" has enough energy to go right through the black tape. In fact, you probably know about some photons that can go through just about anything – X-rays. X-rays are highly energetic (short wavelength) photons and most go right through you. These high energy photons have little hesitation to go through black tape even though the lower energy photons that your eyes can see are blocked. If the energy is high enough, they can go through *anything*. Hence the shielding problem.

Decay

The paper Sally is holding at is a drawing of nuclear decay. This nuclear decay shows Hydrogen 3 (^3_1H or tritium) turning into helium 3 (^3_2He) and ejecting an electron.

Nuclear decay always emits some sort of invisible radiation which can be very dangerous. Properly managed, it can be safe and useful.

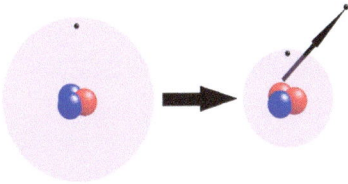

We looked at our sheets.
It was physics alright.
It was one simple question:
What exactly is light?

We were so bored
that we argued for fun!
But Mom wasn't mad
that we got nothing done.

4   5

Illustration 20: The Cat pages 4-5

Illustration 21: Tritium decay

Tritium is hydrogen with two neutrons and one proton in the nucleus. In tritium decay, a neutron becomes a proton and emits an electron as its decay product. The electron can be captured and put to work.

Tritium is sometimes used to illuminate watch dials. The ejected electron lights phosphor the same way older televisions illuminate the screen from beam of electrons hitting phosphor on the screen.

Illustration 22: Tritium illumination

Because the mere mention of the word "nuclear" can have a negative connotation, all references to it were removed from *The Cat in the Box* since it was aimed at young children. It is an important aspect of modern physics, however, and if properly managed can be used safely.

Nuclear decay is a natural process. For example, bananas, granite, and the air under your house all are slightly radioactive. Nuclear decay can be employed in a variety of wonderful uses such as extending the shelf life of milk so it has time to get to third world children.

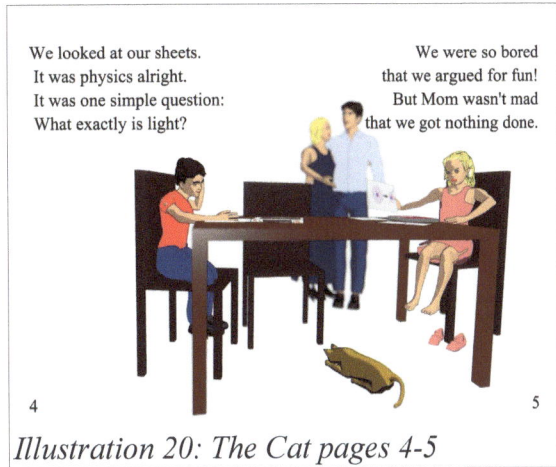 The series logo is carved into the chairs.

Nuclear physics

Some people will tell you how safe nuclear energy is. Others will tell you how dangerous it is. Are the differences in opinion real or are they political?

I often teach a class on batteries. In this class, we have a quote I like:

> "There is no battery so safe it can't be handled dangerously and no battery so dangerous it can't be handled safely."

Nuclear energy is like this too. We have a deep understanding of what types of radiation occur under what conditions and we know how to shield this radiation so that it does not pose a danger. But there are factors such as cost which preclude the proper precautions from being taken all the time and in this case, a danger can be present.

There are enormous gains to be harvested from nuclear reactions. In addition to energy, employing nuclear reactions provide enormous medical benefits and when done safely and responsibly, this technology is a very useful tool. But accidents do happen, and malice does occur, and like anything powerful it can pose dangers. The more powerful something is, the more dangerous it can be when not properly handled.

Nuclear reactions always emit some sort of invisible radiation which can be dangerous if not properly managed. There are many types of radiation. The three most well known are named after the first three letters of the Greek alphabet, alpha (α), beta(β), and gamma *(γ)* radiation. Additionally there is neutron radiation well known for its use in nuclear power plants. In yet another type of nuclear reaction, an electron from the inner shell is kidnapped by the nucleus and is called electron capture. In each type of radiation, a particle is ejected from the nuclear reaction.

In some cases, nuclear radiation can occur by changing the energy or state of the nucleus but without changing its content. An example of this is magnetic relaxation radiation used in Magnetic Resonance Imaging (MRI) machines which we will discuss.

What holds a nucleus together?

The protons in the nucleus are all positively charged. Since like charges repel each other, the electrical forces of the protons make the protons repel each other and try to tear the nucleus part. So why don't all nuclei (nuclei is the plural of nucleus) spontaneously fly apart? There is a competing force called the strong nuclear force that is attractive between all protons and neutrons that holds the nucleus together. These two forces compete and some arrangements are stable and some are unstable. If the electrical repulsion dominates, the nucleus will tear itself apart in a process called nuclear decay. If the attractive nuclear force dominates, the nucleus will be stable. If the electrical force swamps the nuclear force, the nucleus will be short lived. If the competition is a close one, it can take a while for the nucleus to decay. Think about two basketball teams that are closely matched – they'll go into many overtimes.

There is an additional factor to consider with regard to the stability of a nucleus. While a proton by itself is stable, a neutron by itself is not. A neutron by itself, far away from any protons, has a 50% chance of decaying in about ten minutes. If it is sitting very very very close to a proton, it is stable. How close? About 2 femto meters or 0.0000000000001 meters! That sentence actually needs more than three "very"s. But if we try to pack two neutrons with one proton, the single proton cannot quite stabilize both neutrons and the neutrons have a 50% chance of decaying in about 12 years. This is called "tritium" and is the material used in the watch dial discussed above. When one of the neutrons decay, it spews out an electron.

So in addition to the balance between nuclear and electrical forces, you can get decay if there are neutrons without sufficient proton pairing.

The electrical field falls off as the square of the distance from a charged particle just like gravity. The strong nuclear force falls off much much faster. If the particles are really really really close together (say less than 10E-15m, one ten thousandth the size of the smallest atom), the nuclear force dominates. If the particles are only really really close (say more than 10E-15m), the electrical force dominates and the nucleus flies apart. Why the nuclear force is so much stronger at short distances and why it falls off so much faster than the electrical force and gravity is not known at this time - no one has figured this out yet. So if you're looking for a challenging research problem to get your Ph.D…

In summary, if there are too many protons, the electrical repulsion will tear the nucleus apart and it is not stable. If there are two many neutrons, there will not be enough protons to stabilize them and some of the neutrons will decay. It varies from atom to atom, but for most atoms with less than 84 protons there are several stable arrangements with a roughly equal number of protons and neutrons. There are two exceptions: there are no stable arrangements for 43 protons ($_{43}$Tc) or 61 protons($_{61}$Pm). For atoms with more than 83 protons, the nucleus is too big for the nuclear force to dominate and there are no stable arrangements.

Decay and radiation

Alpha decay α

Sometimes in large nuclei the strong force attraction is not strong enough to swamp the electrical repulsion of the protons. A chunk of the nucleus with two protons and two neutrons flies off. We call this particle an alpha particle and the emission of the alpha particle is called alpha radiation. This mostly occurs in elements with more than 82 protons (z>82) and is the most common way for heavy elements to become lighter.

There are no electrons in the alpha particle and it is identical to a $^4_2$He atom with no electrons. When the number of electrons is not equal to the number of protons, we call it an ion rather than an atom. Therefore, alpha radiation could be called helium ion radiation.

An example of alpha decay is polonium-210, or $^{210}_{84}$Po. Reviewing the notation we used for atoms, polonium, element number 84 on the periodic table, has 84 protons and 126 neutrons. 84+126=210. Protons and neutrons each weigh one atomic mass unit, so $^{210}_{84}$Po weighs 210 atomic mass units.

It turns out that a nucleus with 84 protons and 126 neutrons is too large to be stable and the $^{210}_{84}$Po nucleus emits an alpha particle. Since 2 protons and 2 neutrons are in the alpha particle, what's left is 82 protons and 124 neutrons. Atomic number 82 is lead (Pb), so an atom with 82 protons and 124 neutrons is $^{206}_{82}$Pb and we say that polonium-210 ($^{210}_{84}$Po) decays into lead-206 ($^{206}_{82}$Pb), ejecting an alpha particle. If you had a sample of $^{210}_{84}$Po, half of it would decay into $^{206}_{82}$Pb in about 138 days. We call this the half life. So if you had 100 grams of $^{210}_{84}$Po, after 138 days you'd have 50 grams of $^{210}_{84}$Po and after another 138 you'd have 25 grams of $^{210}_{84}$Po. The rest would be $^{206}_{82}$Pb.

The atom, having shed two protons and not shed any electrons, is now unbalanced – it has 82 protons and 84 electrons. We call it a -2 lead ion. The alpha particle, on the other hand, has two protons and no electrons. It is a +2 helium ion.

The alpha particle gets ejected with considerable energy and is traveling fast. As it hits things (such as paper or air), it slows down with each collision.

Alpha Radiation

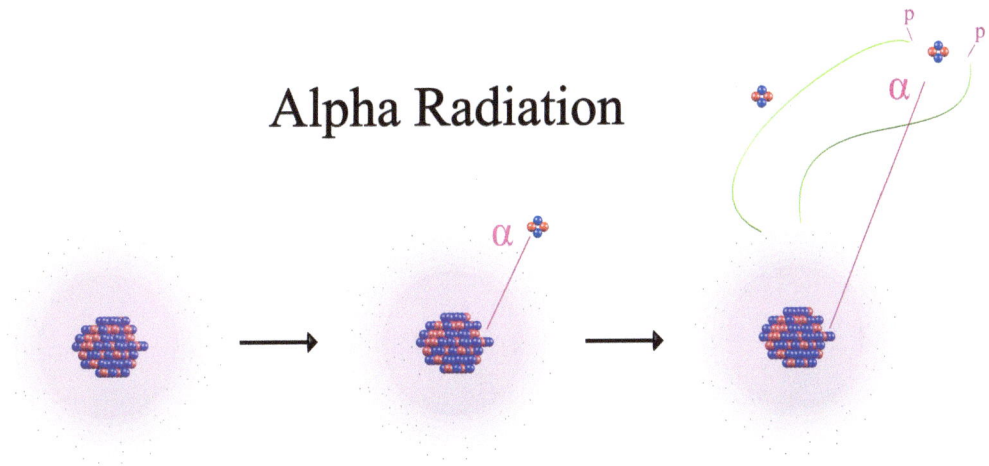

Polonium-210 ($^{210}_{84}$Po)

$^{210}_{84}$Po nucleus being too big for stability, it ejects an alpha particle leaving $^{206}_{82}$Pb.

Because of electrical attraction, two electrons eventually migrate from the Pb to the alpha particle, leaving $^{206}_{82}$Pb and $^{4}_{2}$He. When the electrons join the atom, they give up energy to create two photons.

Illustration 23: Alpha radiation from polonium-210

When the alpha particle has slowed down to the speed of electrons orbiting other atoms, electrical attraction from the alpha particle's protons will attract two electrons from somewhere to form an innocuous, electrically balanced helium atom. Meanwhile, the 84 electrons in the lead ion repel each other more than the 82 protons attract them and two electrons will leave the lead ion, leaving a balanced lead atom. The restoring of electrical balance takes much longer than the nuclear reaction (seconds or minutes compared to nanoseconds). Therefore, it could be said that the two electrons from the lead ion meander and find their way to the helium ion. Because the alpha particle departed with such velocity, it's unlikely that the same two electrons leaving the lead ion are the ones that end up in the helium atom.

When the electrons get attracted and fall into orbits in the helium atom, they lose energy. This energy creates photons which are ejected from the atom. These photons have a lot less energy than what's coming directly from nuclear reactions and are usually in the low end of the x-ray spectrum. Therefore, these are often called soft x-rays.

Alpha particles are easily stopped – skin stops them, as does a piece of paper, and shielding from alpha radiation is relatively easy. But if it is ejected with a lot of energy (i.e., it is moving fast), it can damage the material it hits. The alpha particle ejected from $^{210}_{84}$Po is highly energetic and, although it can be stopped with a piece of paper, if you ingest $^{210}_{84}$Po, it quickly damages your internal organs. It is suspected that in November 2006 Alexander Litvinenko was assassinated by adding $^{210}_{84}$Po to his tea. For more information on the assassination, see http://en.wikipedia.org/wiki/Poisoning_of_Alexander_Litvinenko.

How to save lives with alpha radiation

Americium-241 ($^{241}_{95}$Am), also an alpha emitter, is used in fire and smoke detectors. The alpha particles ionize some of the air and a small electrical current flows through the ionized air.

If there is a hot fire, the hot fire ionizes some of the air around it, heating the electrons enough that they boil off their atoms. These ions and freed electrons spread themselves around the room and some drift into the ion chamber. These ions change the current flowing through the ion chamber and battery. This change is detected and sounds an alarm. Next thing you know you're fanning a towel near the smoke detector to blow those ions away from the detector before your wife wakes up and finds out you forgot to turn off the stove.

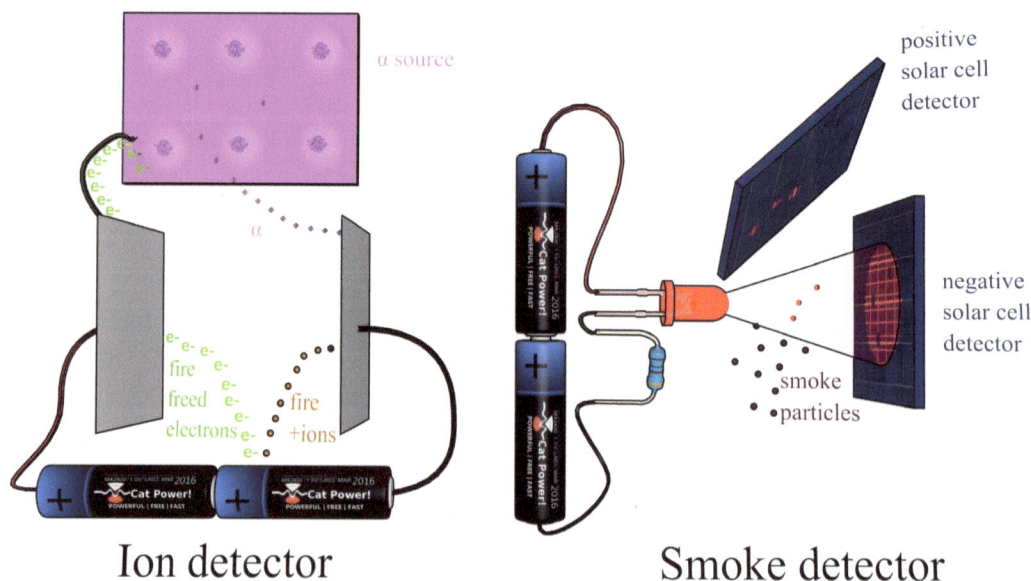

Illustration 24: Ion detector on the left, smoke detector to the right

Technically, this is not a "smoke detector", it is an "ionization detector." For this reason, this detector will not detect slow, smoldering fires, such as a cigarette in a couch. That's why some detectors have both a photo-electric detector to see smoke and an ionization detector to see hot fire. The detector to the right is a smoke detector. The smoke changes the amount of light that hits the light sensors. The light hitting the negative solar cell detector is shadowed by the smoke and the solar cell's output decreases. Some of the light is scattered by the smoke and hits the positive solar cell detector and its output increases. Either technique will work. This change in solar cell output is detected and sounds an alarm.

In some alpha radiation, the alpha particle is ejected with low energy, which translates as a slowly moving alpha particle. In other reactions, though, a lot of energy is released as a fast moving alpha particle. So different energy levels of alpha radiation show up as different speeds of the ejected alpha particle.

Beta decay β

The second type of decay is called Beta (β) decay. In β decay, the balance of protons and neutrons in the nucleus is not stable and a proton becomes a neutron or a neutron becomes a proton. The total number of nucleons in the nucleus does not change.

There are three types of beta decay, β- decay, β+ decay, and electron capture.

β- decay

When a nucleus has too many neutrons, a neutron decays by turning into a proton and ejecting an electron. β- radiation could be called electron radiation.

β+ decay

When a nucleus has too many protons, a proton decays by turning into a neutron and ejecting an positron. β+ radiation could be called positron radiation. When the positron gets ejected, it will slow down as it goes through matter. When it has slowed down to the speed of electrons orbiting other atoms, the positively charged positron will attract an electron. When they collide, the pair annihilates and the energy is used to create a pair of entangled photons.

How to save lives with beta radiation

Suppose we embed a β+ radiation radioactive material in a sugar and add the sugar to someone's bloodstream. We now wait for an hour (probably lying in a thin gown on an unimaginably cold stainless steel table). This sugar has been specifically engineered to be absorbed into a specific tissue of interest. During the hour wait, that tissue absorbs a much higher concentration of the sugar than other parts of the body.

We then put the person into a device called a Positron Emission Tomography imager ("PET" imager). This imager watches for pairs of entangled photons with a pair of detectors. It calculates the equations of the paths (lines) of the photons. By doing the very same math you used to calculate the location of the explosion in the video game, the PET determines the location of the electron/positron annihilation. By recording these for a long time (a minute or so), you can build a complete 3D model of the tissue. It's a *very* clever system.

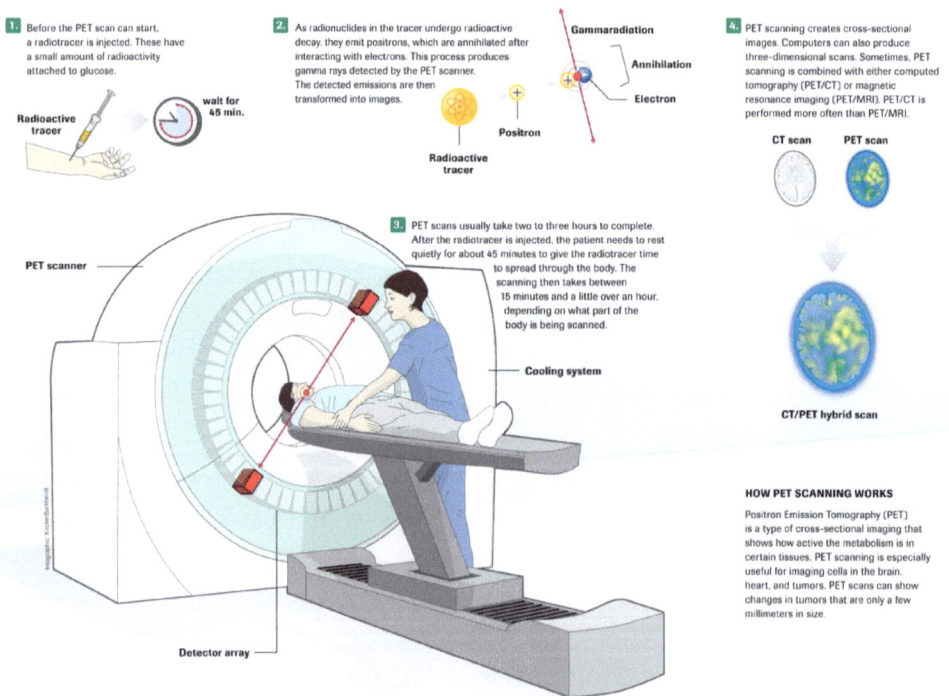

1. Before the PET scan can start, a radiotracer is injected. These have a small amount of radioactivity attached to glucose.

Radioactive tracer

wait for 45 min.

2. As radionuclides in the tracer undergo radioactive decay, they emit positrons, which are annihilated after interacting with electrons. This process produces gamma rays detected by the PET scanner. The detected emissions are then transformed into images.

Radioactive tracer

Gammaradiation

Annihilation

Positron

Electron

4. PET scanning creates cross-sectional images. Computers can also produce three-dimensional scans. Sometimes, PET scanning is combined with either computed tomography (PET/CT) or magnetic resonance imaging (PET/MRI). PET/CT is performed more often than PET/MRI.

CT scan   PET scan

CT/PET hybrid scan

PET scanner

3. PET scans usually take two to three hours to complete. After the radiotracer is injected, the patient needs to rest quietly for about 45 minutes to give the radiotracer time to spread through the body. The scanning then takes between 15 minutes and a little over an hour, depending on what part of the body is being scanned.

Cooling system

HOW PET SCANNING WORKS

Positron Emission Tomography (PET) is a type of cross-sectional imaging that shows how active the metabolism is in certain tissues. PET scanning is especially useful for imaging cells in the brain, heart, and tumors. PET scans can show changes in tumors that are only a few millimeters in size.

Detector array

Illustration 25: How a PET scan works

Depending on the reaction, a beta particle may have considerable energy in the form of high speed. Nevertheless, β radiation can be stopped with a thin metal foil. An example of β radiation is tritium decay which was discussed earlier.

Beta Radiation

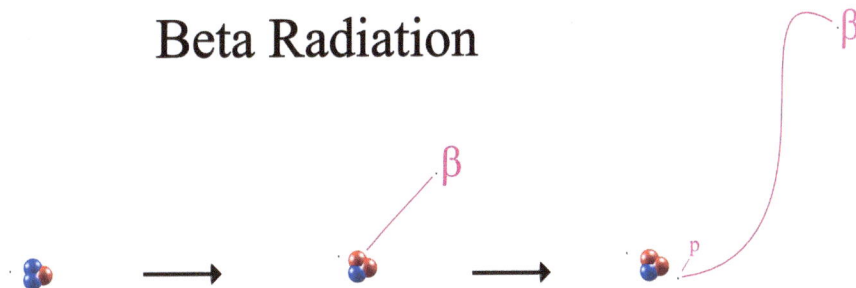

Hydrogen-3 ($^3_1$H or Tritium)

$^3_1$H nucleus has too many neutrons for them all to be stable and it ejects a beta particle turning one of the neutrons into a proton leaving $^3_2$He.

Because of electrical attraction, an electron eventually migrates to the helium ion, leaving a balanced helium atom. The electron goes into an atomic shell and energy is leftover to create a photon.

Illustration 26: Beta radiation from tritium

In addition to using the electron to light a phosphor coating on a watch dial, the electron can be captured and used as an electrical source. In a marketing effort to avoid the negative stigma of "nuclear", commercial versions often call it beta-voltaic or direct β capture. Commercial versions can be found by searching these terms.

Like alpha radiation, some reactions are more energetic than others and the difference in energy shows up as different speeds of the electron (beta particle).

β radiation is the most common type of radiation because it has the lowest energy and therefore can be made from the lowest available energy.

Conservation of charge

One of the constants of the universe is charge: the amount of charge in any isolated system (including the universe) never changes. So if you somehow manage to create a negative charge (like ejecting a β-particle), a positive charge will also be created. We say "charge is conserved". Note that in tritium decay, a neutral neutron becomes a positive proton, balancing the charge from the newly created negative β-particle. As far as we know, charge is always conserved.

Gamma radiation γ

In the third type of radiation, gamma (γ) , the ejected particle is a photon. Sometimes a decay gives up more energy than is needed to eject the alpha or beta particle and that energy creates a photon. Suppose, for example, that an alpha decay has occurred in $^{210}_{84}$Po. Most of the time, the alpha particle is made from protons and neutrons on the outside of the nucleus and once the alpha particle is ejected, there is no more nuclear activity. But once in a while, about one in 100,000, creation of the alpha particle leaves a hole in the lead's nucleus. The protons and neutrons then rearrange themselves in a lower energy state and fill in the gap. The excess energy forms a photon which gets ejected and is called a gamma ray. This excess energy nuclear is much higher than the electron shell energy left over that formed the soft x-ray discussed in the alpha radiation section. This is called a hard x-ray.

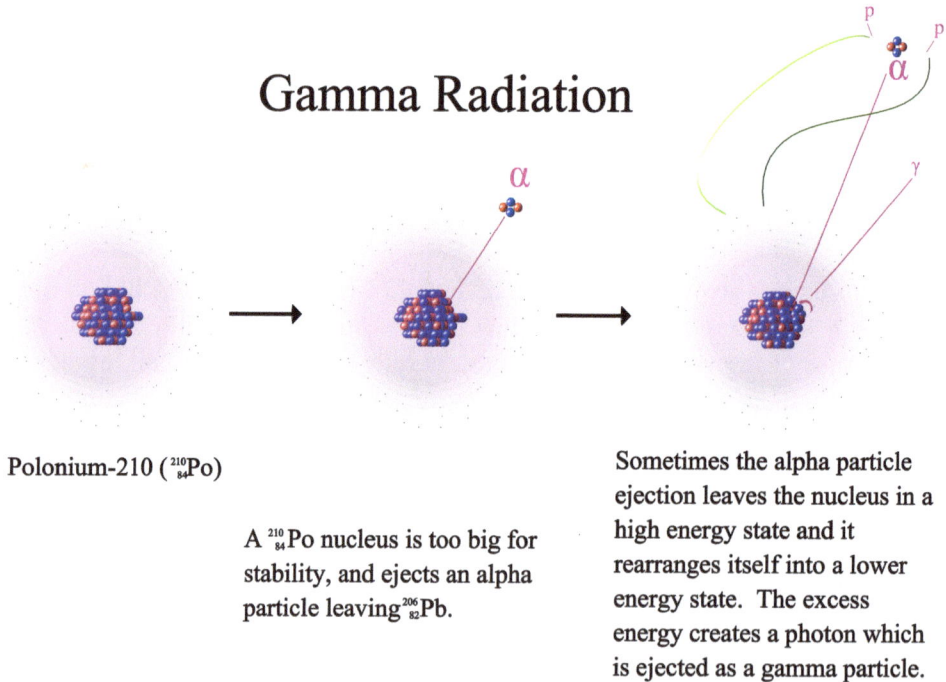

Gamma Radiation

Polonium-210 ($^{210}_{84}$Po)

A $^{210}_{84}$Po nucleus is too big for stability, and ejects an alpha particle leaving $^{206}_{82}$Pb.

Sometimes the alpha particle ejection leaves the nucleus in a high energy state and it rearranges itself into a lower energy state. The excess energy creates a photon which is ejected as a gamma particle.

Illustration 27: Occasional gamma radiation from polonium-210

Although this is a rare occurrence in Po, there are many decays that often or always emit gamma radiation.

Neutron emission

In neutron emission, the ejected particle is a neutron. This radiation is very important in nuclear power plants because the radiated neutron bombards the fuel causing further nuclear fission (fission is when the nucleus splits apart). This further nuclear fission creates more neutrons and the process goes faster and faster in what's called a chain reaction. If this keeps up, the chain reaction can get out of hand and we cannot control it. When this happens, it is called a meltdown. It is very important, then, to control the rate of reaction. How do we do this?

Imagine starting a small fire in a large pile of dry wood and wanting to keep the fire a certain, small size. The propagating mechanism in a fire is the spread of heat. To control the fire, you could carefully add small amounts of water at the right time in the right place to control the propagation. In this way, you control the rate of the small fire spreading by regulating its size with water which absorbs heat, the propagating mechanism of a fire.

In fission reactors, neutrons are the propagating mechanism. To control the reaction, we insert a neutron absorbing material (often cadmium or boron). This is analogous to adding water to the small fire. This material is made in the shape of a rod. The material is carefully raised and lowered into the nuclear material to control the reaction. These are called control rods. As described by Enrico Fermi and Richard Garwin, it helps that about 0.5% of the neutrons from fission are delayed by about 10 seconds.

An additional concern is that neutron radiation can make non-radioactive material radioactive. A neutron can bombard a stable atom and change the content or energy of the stable atom, making it no longer stable.

Electron capture

In electron capture, the ejected particle is a small, uncharged particle called a neutrino. What happens is this: an electron from the inner shell of the atom gets attracted to and combines with a proton in the nucleus. The combination of a proton and electron forms a neutron which stays in the nucleus. There is some energy left over and gets used to create yet another particle called neutrino which gets emitted. An outer shell electron then replaces the inner electron, emitting a photon in the form of a soft x-ray.

An example of electron capture is ^7_4Be (beryllium-7) which turns into ^7_3Li (lithium-7).

Resonance and Magnetic Relaxation Radiation

Many things have a natural frequency called its resonant frequency. Some examples that you might have in your house are a guitar string (on a guitar, not one sitting in its package), a wine glass, and a tuning fork. Take one of these and pluck or tap it to hear its resonant frequency. It not only likes to send out this frequency, it likes to take it in! Do this: pluck or tap it and listen carefully. Now put your finger on it to quiet it. Remove your finger and sing that same note loudly for about a second. Now stop and listen. Did some of the energy from your voice get transferred? Is it making noise at that same note?. It won't be very loud, so you may have to listen carefully in a quiet room. Now try this by singing a different note. It doesn't work! You have to sing at the object's resonant frequency for it to work.

This is called resonant power transfer. You may have seen an opera singer cracking a glass by singing very loudly. The opera singer sings at the glass's resonant frequency adding energy to the glass until it cannot vibrate that hard and it shatters. Pretty neat trick.

Many things have a resonant frequency including the nuclei of atoms. When you put an atom into a magnetic field oscillating (turning on and off) at the nucleus' resonant frequency, the nucleus absorbs that energy. If you then turn off the magnetic field, the nucleus gives off that energy in the form of photons just like the guitar string or glass gave the energy it absorbed back in the form of sound. Like the guitar string and glass, it only does this at the resonant frequency.

Different nuclei have different resonant frequencies. Suppose you have a sample and you want to know what's in it. By starting with a low frequency and slowly increasing the frequency until you are at a frequency higher than any atom nuclei, you can look for resonances and see what atoms are in the sample. Clever, eh?

This machine is called a Nuclear Magnetic Resonance (NMR) machine.

If you think that is clever – check this out: instead of changing the frequency, set it to the resonance of a hydrogen atom. Remember the video game where we can find the location of the explosion and how that same math can be used to image the photons from a PET imager?

An imager similar to the one described above in the PET imager detects these photons, uses the same math you used in the video game example at the beginning of this book, and determines the location of the hydrogen atoms. Although there is hydrogen in many things, inside your body it is mostly in water. Therefore, what you end up imaging is the density of water in your body. Since your organs have a different density of water than things surrounding them, you can get a 3D image of an organ. While an X-ray shows hard things such as bones, this machine will show soft tissue – something that an X-ray will not show.

What's in a name?

Sometimes *everything* is in a name. This imaging Nuclear Magnetic Resonance (NMR) machine was invented to provide imaging of soft tissue inside the human body without opening the skin. It was a commercial failure because no one was willing to put themselves into a "nuclear machine." GE then renamed the machine Magnetic Resonance Imaging (MRI) and it has enjoyed tremendous commercial success.

Nuclear power generation

The products from nuclear power plants emit high energy alpha, beta, gamma, and neutron radiation. Gamma and neutron radiation require a large amount of shielding and pose a large health issue. Further, their half life is tens of thousands of years, so you have to shield it and keep it away from people for a very long time.

A major promise of the National Ignition Facility (NIF) in Livermore California is nuclear reactions that don't produce as much long life radioactive waste. NIF is exploring fusion – the opposite of fission we have been discussing. In fission, you're breaking a large nucleus apart. In fusion, you're taking two small nuclei and fusing them together. At NIF, you start with hydrogen, make helium, and neutrons are ejected. While the neutrons are relatively high energy, they are short lived and with proper shielding the radioactive waste problem is reduced. Another promise of NIF is inexpensive energy. This is an exciting area of research which could be a true game changer.

There are many possible fusion reactions but none have proven to be practical yet. Fusion is not a panacea, but is better than fission. A fusion reactor can't "blow up", because the fuel needs carefully controlled conditions to react (so much so that we can't get to work well yet!) Any damage to the reactor causes it to stop. The only remaining danger is the much smaller radioactive material present. One particularly attractive reaction is fusing helium-3 ($^3_2$He). The Wikipedia article on helium-3 nicely describes this process.

Can I electrically control decay rate?

After reading this chapter, one of my friends asked, "If the balance between electrical forces and nuclear forces determine nucleus stability, then can I control the rate of decay with an electric field?" That is a very good question.

Theoretically, the answer is yes. Subjecting the nucleus to an external electric field could upset the delicate balance between the electric field and nuclear force. But there's a problem. Since the neutrons and protons in the nucleus are *very very very* close together, their electric field is *very very very* strong.

The only way to get an electric field is to separate protons and electrons. These external charges are going to be a more than a billion times as far from a proton in the nucleus than another proton in the nucleus.

The strength of an electric field falls off as the square of the distance. So if the external protons and electrons are a billion times as far away, we need a billion times a billion of them to influence the effect if a single proton in the nucleus. How many is that? If you had two pounds of nuclei, you'd need as many protons and electrons as the earth has to influence decay rate!

Even if you could manage to get the mass, and you could manage to strip off all those electrons, the attractive force between the protons and electrons would crush any structure you made to keep them separated.

So, practically speaking, no you cannot in practice use an electric field to control decay rate under the 'ordinary' conditions we can produce. But in stars or outer space near a star, the situation is different, and nuclear decay can be significantly affected. For example, in a neutron star, crushing gravity can squeeze electrons into protons, making stable neutrons, and beryllium-7($^7_4$Be), unstable under 'ordinary' conditions because of electron capture, is stable in space near a star where all electrons are stripped.

There are a few rare cases on earth where external influence affects nuclear decay rates. The rate of electron capture of $^7_4$Be depends, by a few percent, on its chemical environment because in some chemical compounds, the electron density near the nucleus is affected. So if you form certain compounds with $^7_4$Be, its decay rate can be affected by a few percent by its chemical state. If you tried to make such an effect with an electric field, the effect would be orders of magnitude smaller because the largest laboratory fields are orders of magnitude smaller than the internal fields of atoms and molecules.

Is light pieces or waves?

Is light made up of particles like sand or is it a wave like water in the ocean? This is a debate which has been going on for a *long* time.

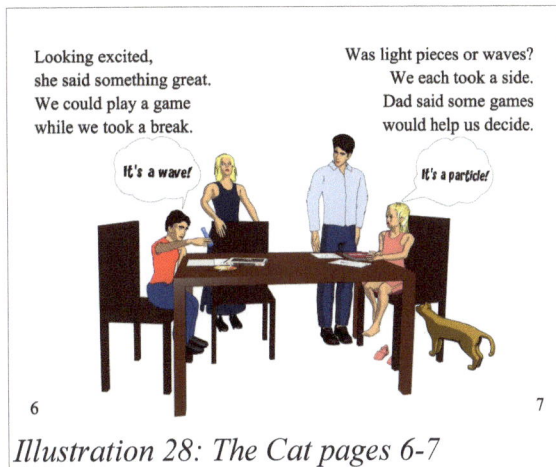

Illustration 28: The Cat pages 6-7

In 1666 Isaac Newton concluded light was made of particles based on his observation that light traveled in straight lines.

Noting that particles follow his three laws of motion and travel in straight lines, Newton observed that light seemed to follow those laws. His reasoning was that light waves appear to travel only in straight lines while ocean and sound waves seem to bend around obstacles. This is very nicely described in Chad Orzel's "How to teach physics to your dogs," explaining that's why "... a dog in the dining room can hear a potato chip hitting the kitchen floor even though she can't see it."

In 1678 Christiaan Huygens concluded that light was a wave. He based this on something called refraction and diffraction, which we'll talk about shortly. For details, take a look at The Project Gutenberg eBook, *Treatise on Light*, by Christiaan Huygens.

The theory of light as a wave was further substantiated with interference patterns which we will talk about in detail.

In 1905 Albert Einstein concluded light was particles, based on the photoelectric effect, observing that light knocked electrons out of a metal in discrete chunks.

To see why there were so many disagreements about this seemingly simple question, we shall perform some experiments.

We will start with experiments with sand which is clearly made of particles and water waves which are clearly waves. That way, we can learn the characteristics of particles versus the characteristics of waves. Then we can see what happens when we repeat the experiments with light and see whether it follows the characteristics of particles or the characteristics of waves.

The flashlight is illuminating the series logo.

Particle experiments

Particles are things that come in clumps, or discrete pieces. You can count the number. You can describe the location and weight of a particle.

This is not the case with waves – asking exactly where a wave is or how much it weighs doesn't make sense.

Illustration 29: The Cat pages 8-9

We begin with some experiments with rice. Rice is obviously a particle – if I give you some rice you can tell me how many grains there are and tell me where each grain is. Through these simple experiments we will learn some of the characteristics of particles. Next we will do some experiments with sound and water waves, which are obviously not particles – if I turn on the radio, the questions of exactly where the sound is and how much it weighs don't make sense. You can tell me where the sounds waves are coming from (the radio), but you can't tell me exactly where the sound wave is because it is spread out. You can tell me how much the radio weighs, but you can't tell me how much the wave weighs. The radio is a particle, the sound is a wave. Particles are countable. There is one radio, but how many sound waves are there?

Tim is dropping rice grains to see what a few particles look like when they are dropped. Sally is learning that a picture taken in very low light levels has characteristics similar to when you drop a small amount of rice.

Then we'll repeat the experiments with light and see if light follows the characteristics of particles, coming in little bundles and having a specific location.

The series logo is on the left side of the table under the ripple tank.

Particles exhibit the characteristic that they come in clumps, are countable, are in a definite location, and have random distributions when there aren't very many of them and you are looking very closely at them. In this experiment, we will take rice and drop it on the table from about a foot above the table. We will look at a small number of particles and then look at a small number of photons of light to see if the result is similar.

Exercise 1: Drop a small amount of rice (say 100 grains) onto the table and look at the pattern.

You don't have to use rice. You can use any particle – you can use unpopped popcorn, for example. When you do this experiment, you'll probably have the same problem that I did – the particles bounce all over the place. It's difficult to see the pattern where they landed. So what I did was drop rice into some water so it didn't bounce away. Here's what it looked like when I did it.

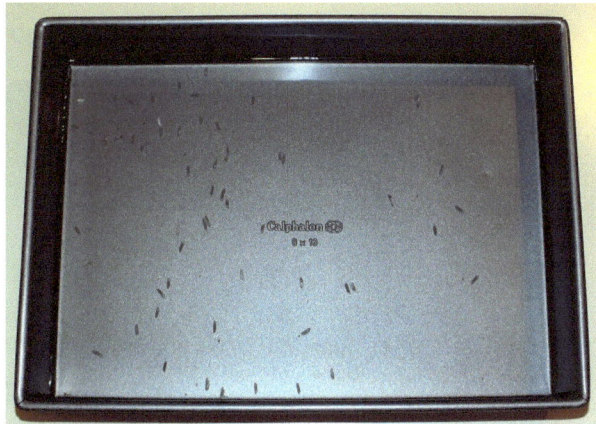

Illustration 30: Particle test for rice

You'll notice that the rice comes in clumps of a certain size – you don't find a nice range of sizes. Each clump starts and ends in a crisp fashion, not slowly spread out like a fuzzy picture or like a wave.

Exercise 2: Use a camera to take a picture in very low light level and zoom in on the picture.

Compare the two and note the similarities and differences.

Here is the picture I took with all the lights out in my bathroom which doesn't have any windows. It was a 10 second exposure:

Illustration 31: A really dark picture

Here is is what part of the picture looks like when zoomed way in:

Illustration 32: Dark picture zoomed in

Notice that the light came in bundles, is countable, and is in specific locations.

Is this really what we're seeing?

Caveat: In reality, the pattern that is being recorded in your camera may or may not actually be due to the particle nature of light because there are a number of other phenomena that can give similar looking pictures. When taking pictures at such a low level of light, it is difficult to control which one dominates. For example, thermal noise within the camera's light sensor can introduce noise which looks remarkably similar to the particle phenomena pattern. It takes a lot of homework and patience to do this particular experiment right. Nevertheless, when you "do it right" as Einstein did in 1905 (see *The miracle year of 1905* section) and in every test since then, the results show this pattern.

I bring this detail up to illustrate something about being a scientist. In science, it is not enough to refrain from lying. In science, a more brutal sense of honesty is required which obligates us to include mention of any possibility we know of that could cast doubt upon the information we are presenting.

This unusual level of integrity is necessary for several reasons. The most important two are the avoidance of misleading the reader and the avoidance of fooling oneself. It is so very easy to misuse presentations which look like science but are not. When presented correctly, science does not mislead. When the scientific method is properly followed, the results are the best methods humans have developed to form conclusions.

Unfortunately, the name science is sometimes used when the scientific method is not used and the term scientific is devalued. I often hear advertisements say "scientifically proven to..." This is a dead giveaway because no scientific theory is *ever* proven. See the section *When is a scientific theory proven?*. At best, it has withstood all challenges presented to it so far.

The scientific method: how do we know what we know?

For a long time, it was thought that atoms were not further divisible and represented elementary particles. Confidence was so high that these particles were called atoms from the Greek *atomos* which means indivisible. So "atom" literally means "can't be divided." Later it turned out that the atom could be taken apart because electrons could be stripped off. When all the electrons are stripped off, what's left is a nucleus consisting of one or more protons and neutrons which for a long time were thought to be elementary particles. It was later discovered that the nucleus could be broken down into protons and neutrons, and that these, too, could be broken further into smaller components called quarks. The list of what we consider to be elementary particles today is very long – well over a dozen. The electron and the photon are at this time (5:32 AM EST July 20, 2014) thought to be elementary particles and no one has been able to break them apart into smaller components. It has been time tested against many challenges and is now widely accepted. But tomorrow, while highly unanticipated, someone could prove this wrong.

How would this be done? It doesn't get overturned by someone just making the claim that it is wrong! The scientific method requires making a hypotheses, deriving predictions from them as logical consequences, and then carrying out carefully designed and controlled experiments based on those predictions to determine whether the hypothesis was correct. We'll talk about controls in detail in the *Experiment Controls* section. The experiments are then repeated by different people. If and only if the results are repeatable, the new hypothesis will be accepted in the scientific community. But is is never proven – it could always be overturned with a new experiment.

Wave Experiments

A wave is a moving disturbance in a material. For example, when someone dives off the diving board in a swimming pool, she disturbs the water and a pattern of waves travel from her in all directions. The waves cover a lot of space and change over time. Although the wave travels, the material in which the wave is traveling does not travel along with the wave. When the diver goes into the pool, water moves up and down, but it does not travel toward the shallow end of the pool even though the wave travels there and bounces off the wall.

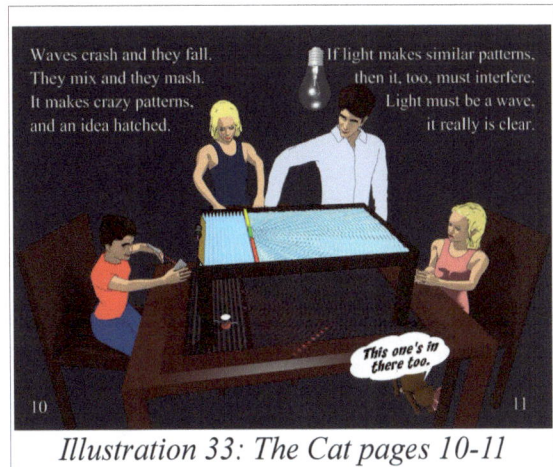

Illustration 33: The Cat pages 10-11

If there's a beach ball sitting in the middle of the pool, the waves will lift it up and down, but it won't move it sideways – unless there's some other force like wind or children, the beach ball will stay in the middle of the pool.

Waves do move other things (hence shells on the beach), but the medium itself – the water – does not get moved in that way.

Waves have recognizable characteristics. Waves the bounce off surfaces and change directions (1). They travel faster in some materials than others (2). Waves can bend around small objects (3). And waves spread out if they start out small or are traveling through a small slit(4). If two waves travel to the same place, the material is influenced by both waves in a way that is called interference(5).

Combining (4) and (5) forms the famous double slit experiment (6) and is a common test to see if something can be described as a wave by observing the results as wave travels through two closely spaced small slits. This is what's shown on the wave table in the picture at the top of this page. The double slit experiment is what Christiaan Huygens based his conclusion on that light was a wave in the 1670's.

First, we'll do the classic double slit experiment with water waves in a ripple tank as shown on page 10 and 11 of *The Cat in the Box*. This will teach is some of the characteristics of waves. We'll then do some more experiments with water, rice, sound, and light.

The series logo is on the left side of the table under the ripple tank.

Double Slit Experiments

We will not be demanding too many experiments, but the set of double slit experiments is the quintessential set to give you experience with wave interference, a corner stone of quantum mechanics. If there is a single set of experiments you should do as a child to become a modern physicist, this is it!

We begin by sending waves through a pair of narrow slits and observing their behavior. We'll do similar experiments with particles and observe their behavior. We'll also look at the results with sound. We'll then send light through double slits and see if it acts as waves through the double slits. The results of the double slit experiment is what forced physicists to resort to quantum physics. This experiment, in various forms, contains the heart and soul of quantum physics. In fact, all the mysteries and intrigue of quantum physics can be shown in this one set of experiments.

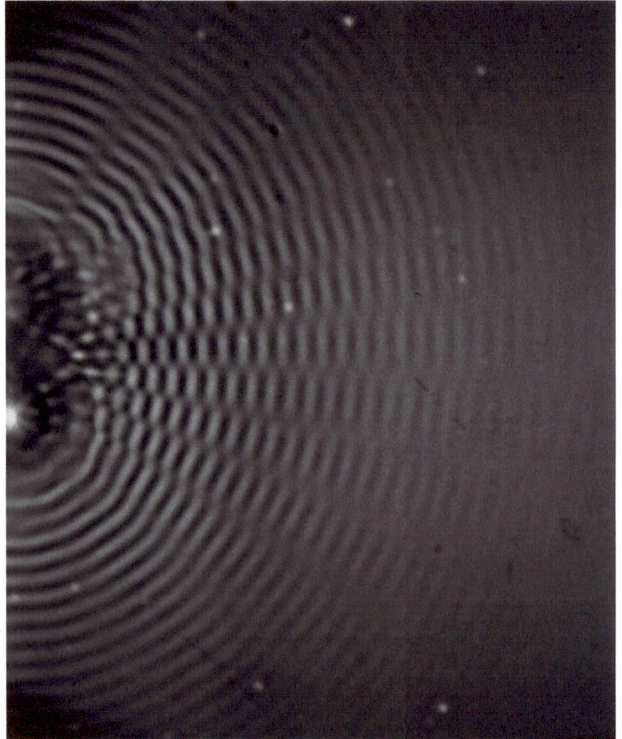

Illustration 34: Double slit in a ripple tank

This works for any type of wave including light, water, and sound. It does not work for particles. The wave table above shows how it looks with water in something called a ripple tank. We will do the slit experiments with water waves where you can see this interference pattern. We will also do it with sound so you can hear the interference pattern. We will do slit experiments with rice where you will see no such effects since rice is made of particles and not waves. Then we'll do the slit experiments with light to see if it exhibits these wave characteristics.

We'll begin with the double slit experiment using water waves in what's called a ripple tank. You can purchase a ripple tank off the Internet. The prices start at about $50 and go up from there. But a much better approach is to build one. There are plenty of plans for them floating around the Internet. Building the setup is a fun weekend project to do with your child. But whether you choose to build or buy, doing these experiments is vital to preparing a person for modern physics.

We will then do a variant of the experiment. Rather than using two slits, we will use two identical sources with each source at the location of one of the slits. The reason for doing two sources rather than two slits is because with today's technology it's easier to get very good results with a dual source than with the double slits, especially for sound. We'll start with water waves and do both a double slit and a double source and verify that the results are the same.

We'll begin with the double slit experiment with water waves as shown on page 10 and 11 of *The Cat in the Box*.

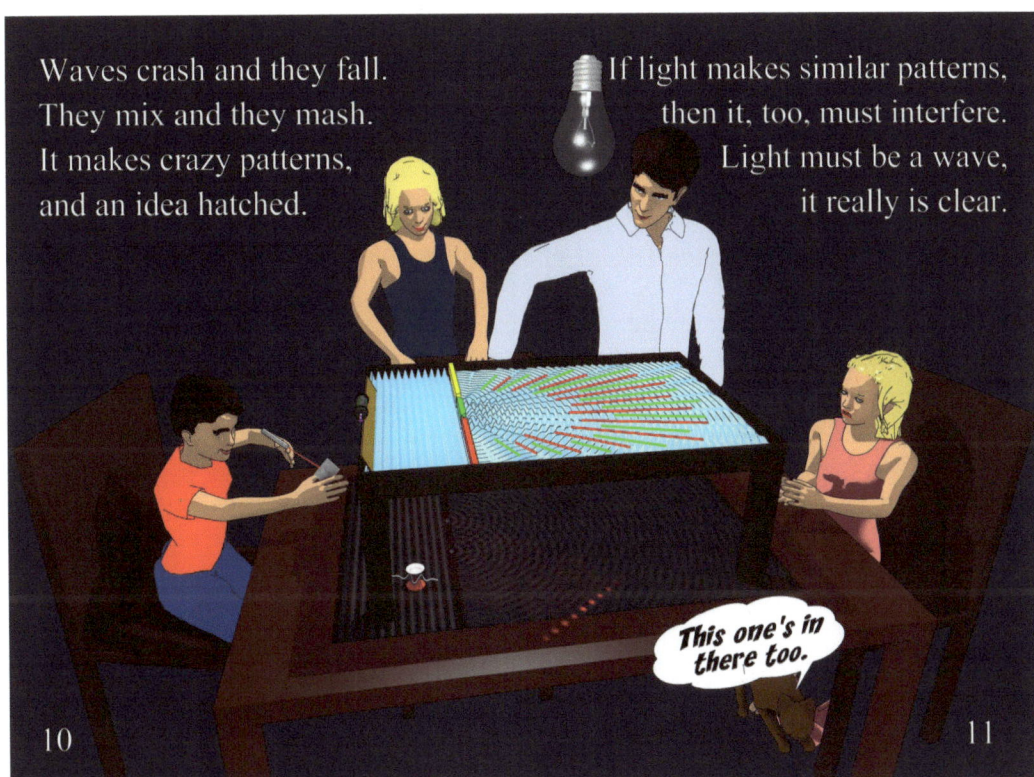

Waves crash and they fall.
They mix and they mash.
It makes crazy patterns,
and an idea hatched.

If light makes similar patterns,
then it, too, must interfere.
Light must be a wave,
it really is clear.

This one's in there too.

10   11

Illustration 35: The Cat pages 10-11 with crests and nulls shown

When you have two identical wave sources, whether they are two separate sources or created by one source and two slits, you get a characteristic pattern called an interference pattern. You can see the interference pattern on the right side of the wave table where the crests from the left slit line up with the crests from the right slit to form tall waves and where the crests from the left slit line up with the troughs from the right slit canceling out the waves, leaving relatively calm water.

Where the crests coincide the crests we call it constructive interference. When a crests coincide with troughs, we call it destructive interference. Shown in Illustration 35 is the experiment with green lines drawn where there is constructive interference and red lines where there is destructive interference. This interference pattern does not happen with particles.

Experiment 1 – Playing with waves

To test the apparatus, turn it on and *play* (Playing pays off! See the last section in this book, *Having fun*). Watch the waves travel from left to right. What happens to the waves if you put something much larger than a wave in the water? What happens if you put something smaller than a wave into the water?

You'll notice that the waves can bend around small objects. It depends on the relative size of the object compared to the distance between two crests of the wave, called the wavelength. This doesn't work just with water waves, it works with all waves including ocean waves, light, and sound. This bending is called diffraction.

The wavelength λ of sound that a child can hear ranges from about half an inch for high frequencies to about 100 feet for low frequencies. The wavelengths for a 88 key piano range from about 3 inches to about 36 feet. Diffraction explains why you can hear someone speak when they are around the corner – the sound waves bend around the turn between the hallway and walls. Some of the sound you hear is bouncing off the walls, but some of it is diffraction. This can be proven by covering the walls with sound absorbing material so that there are no reflections and observing that the sound still makes it around the corner. Since longer wavelengths diffract better, this explains why noises get "muffled" as they go around corners – low frequencies diffract better than high frequencies.

What happens if you put a barrier all the way across the table and block the waves? You'll see waves that are no longer traveling across the table. Instead, you find waves going up and down but not moving across the table. These are called standing waves and are the result of wave reflections and interference.

We see that in this arrangement standing waves in some places and no waves at all in other places. This is an important consideration in microwave oven design so that you do not heat the food in one place and leave it cold in another. The solution to this problem is twofold: 1) Microwave ovens have a moving reflector in them so that the standing waves move around. This is called the microwave stirrer. 2) Many microwave ovens rotate the food. Now you understand why this is necessary.

We'll explore these interesting wave phenomena in an upcoming book on waves.

"Behavior"

Many people use the word "behavior" to describe what something does and it is common to come across an experiment that "demonstrates how photons behave: when do they behave as particle and when do they behave as a wave"? They will talk about the photon going back in time to and "deciding what path to take." It is important to understand, however, that the photon is not a sentient being and does not make decisions to behave the way people do. The photon is not alive, it does not know, understand, or act any more than a baseball "knows" to curve when it goes over the plate. But we're stuck with our language and words like "behave" are often employed.

Experiment 2 – Double Slit Experiment with Waves

We now do the famous double slit experiment for waves with the arrangement shown in Illustration 33. Waves refract out of both slits and start spreading out. Then an interesting thing happens – the waves start combining. When they combine, in some places the wave crests from the left slit are lined up with the crests from the right slit. Where this happens the crests are twice as big and we say the waves are in phase and add constructively. In other places, the crests from the left slit are lined up with the troughs from the right slit. Where this happens the waves cancel each other out and there's no wave left. We say the waves are out of phase and add destructively. The resulting pattern of waves is called an interference pattern. This pattern is a result of waves adding.

Diffraction and interference patterns are tell tale signs of waves. We can use the presence or absence of diffraction and interference patterns to determine whether something can be described as a wave: when we observe diffraction and interference patterns, we find that wave mathematics can be used to describe and predict the results.

Illustration 36: Double slit arrangement

Illustration 36 shows the setup I used for the ripple tank double slit experiment. Unlike the setup on pages 10 and 11, this ripple tank has a mirror and screen under the tank. If you just want to see the results with your eyes, you don't need the mirror and screen. But if you want to take pictures, this makes it much easier.

Here is the result from my ripple tank double slit experiment. You can clearly see the interference pattern. When you do this experiment you may find it hard to see the waves in the water. If you shine a light from above, the shadows under the table will be easy to see. This is a picture of the shadows, not of the water. You can also put a light under the table and see the shadows in the ceiling. For this reason, make sure you use a transparent material for the bottom of the tank.

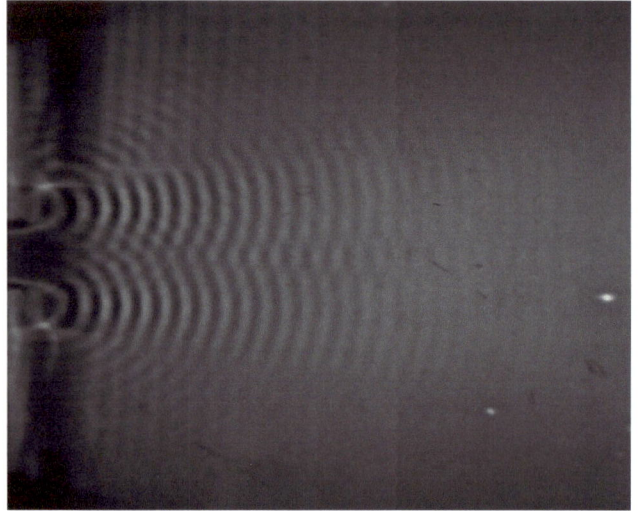

Illustration 37: Double slit ripple tank results

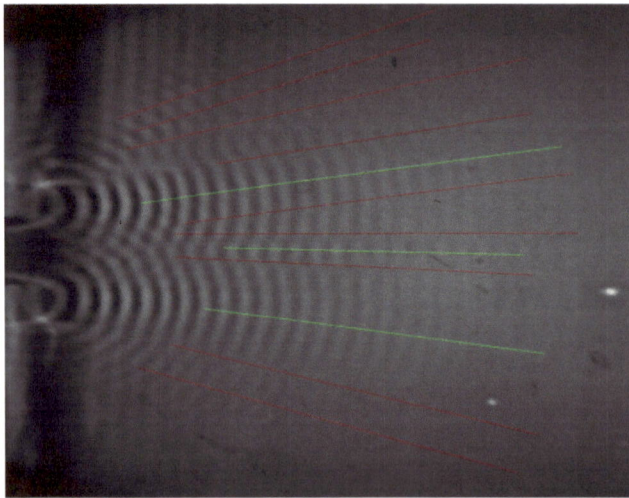

Illustration 38: Crests and nulls in a ripple tank

It is easier to see in the live experiment when things are moving than it is in a still picture, but what's most important here are the areas of constructive interference. To make them easier to see, I've drawn red lines where the destructive interference is greatest and green lines where constructive interference is greatest. As noted above, when you perform this experiment with a ripple tank, you'll easily see motion. Where the red lines are, the water is calm and almost nothing is happening. Where the lines are green is where the most action is and the waves are strongest.

Experiment 3 – Double source experiment with water

As mentioned earlier, with today's technology, a double source experiment gives better results than a double slit experiment because the waves from a point source are more circular than those from a slit.

Here is the setup I used for the ripple tank double source experiment. Instead of hitting two slits with a plane wave, we create the two point source waves to start with. Each source acts as a point source for a circular wave. The two waves travel outward from their source and when they overlap, you can see an interference pattern.

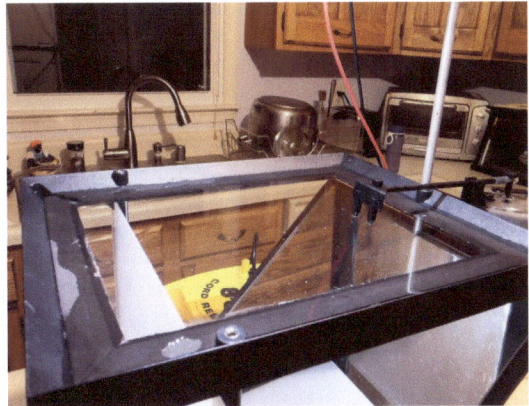

Illustration 39: Double source setup in ripple tank

This is a close-up shot of the double source in action. If you look closely, you can see the waves near the sources. But if you look a couple of inches away, the waves have gone down in amplitude and you can no longer see them. Their shadow, however, can be seen throughout the table. This is why we use a light above and look at the shadows below rather than looking at the waves directly.

Illustration 40: Close up of ripple double source

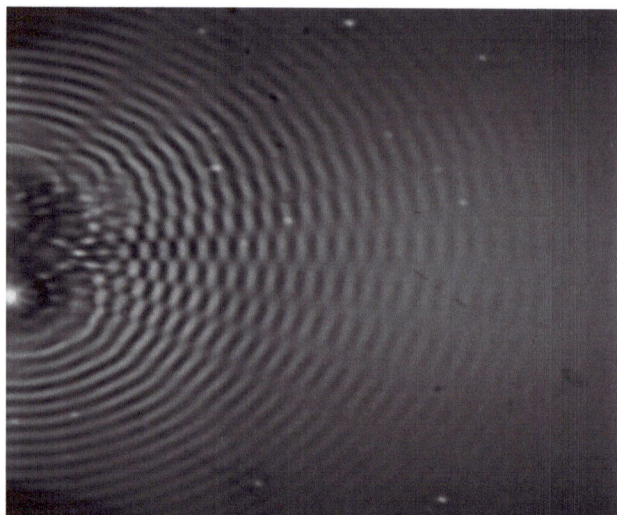

Illustration 41: Results of ripple tank double source

Here is the result from my ripple tank double source experiment. You can clearly see the interference pattern. If you compare this to the double slit result Illustration 37, you'll notice that they are basically the same. This result has the nulls closer together because the two sources were further apart than the slits. Also, the pattern is easier to see because the point sources give better circular waves than double slits.

Experiment 4 – Double source experiment with particles

We now try the double source experiment for particles to see if they exhibit the wave characteristics of interference. For particles, let's use something that is definitely made out of particles such as rice or unpopped popcorn. I used rice.

What I did was to take a piece of posterboard and cut holes in it as shown in Illustration 42. I folded it and put in on cereal boxes to hold it above the tray. I used funnels to make it easy to drop the rice into the holes. I put water in the tray so that the rice didn't bounce around and mess up the pattern.

Illustration 42: Double source experiment with rice

Here is the result – two separate piles of particles with no interference pattern. The rice went through the funnels and fell according to Newton's laws of motion without interfering with other grains of rice. No surprise here.

Illustration 43: Double source rice results

Experiment 5 – Double source experiment with sound

We will now do the interference experiments using sound. While we could do it with slits, it is easier to use two speakers to generate the waves which will interfere with each other – a double source experiment.

Let's use one foot waves. To calculate the frequency we need, we use the equation

$$f = \frac{v}{\lambda}$$

Where
    f is the frequency of the sound in Hertz, or wave cycles per second
    v is the velocity of sound through air, which is 1000 feet per second
    λ is the wavelength of the sound in feet, we want 1 foot

$$f = \frac{v}{\lambda} = \frac{1000\,\frac{feet}{second}}{1\,foot} = 1000\,\frac{cycles}{second} = 1000\,Hertz = 1KHz$$

We will use a sound source of 1KHz (pronounced "one keeloe hurts"). We are using a component of the metric system where K stands for a thousand. Further, the number of cycles per second is named after Heinreich Hertz and is usually abbreviated Hz. So 1000 cycles per second, or one thousand Hertz is written 1KHz.

What does 1KHz sound like? If you've ever heard a teenager scream on a roller coaster, that's it – about two octaves above Middle C.

There are many ways to generate 1Khz. The first test I did used an expensive function generator. Then I did it with an inexpensive function generator. Then I generated a wave using audacity, an open source audio editing software package. Then I searched youtube for ("1KHz sine wave") and tried the first one that popped up. All of these techniques worked. The youtube video played through a cell phone with a pair of computer speakers plugged into the earphone jack was used in the set-up for the results I'll show here.

I tested a variety of different speakers over a large range in price. In general, high quality speakers work better than lower quality speakers, but even the least expensive pair of speakers worked adequately well. I did find that the speakers needed to be well matched – if I used one speaker from a stand alone stereo system and used one from a computer monitor I was not able to get good results. A medium quality set of computer speakers plugged into a cell phone was used for the results I'll show here.

Here is the setup we used:

Illustration 44: Double source setup for sound waves

Notice that the speakers are pointed upwards at about 45 degrees. This is to avoid a bounce off the ground. If you point them in the usual way, there is an additional path – from the speakers, bouncing off the ground, and then back into your ear. This bounce is enough to mess the test up completely. This is an important concept worth spending some time on.

Challenges from multipath

Shown in Illustration 45 is the sound coming out of two speakers. Notice that the sound doesn't go in just one direction, and that it doesn't go out in all directions either. Instead, the sound comes out in a cone shape. For some speakers the cone is a large angle. That speaker has a pattern called an omnidirectional pattern. Other speakers have a small angle cone and are called unidirectional speakers. The speaker on Tim's right is a unidirectional speaker and has its cone shown in red. Notice that none of the waves make it from the speaker to Tim's ear. Unless one of the red waves bounces off something and go back to his ear, he will not hear anything out of that speaker.

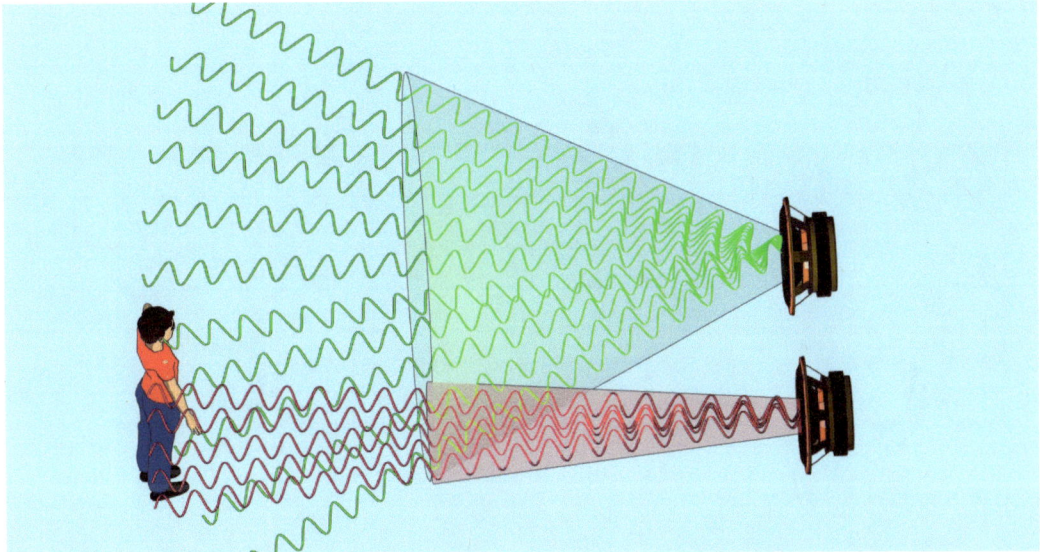

Illustration 45: Omni directional speakers versus unidirectional speakers

The speaker on Tim's left has its cone shown in green. Notice that only one of the waves is going from the speaker to Tim's ear. This wave is shown in green in *Illustration 46* and is a direct path from the speaker to Tim's ear. All the other waves go somewhere else.

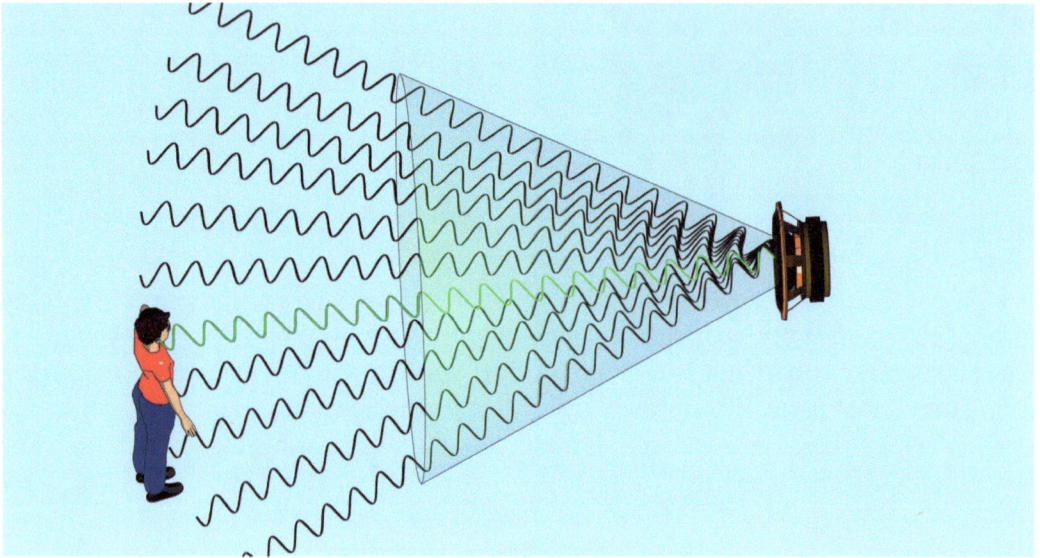

Illustration 46: Speaker pattern

So what Tim really hears is this:

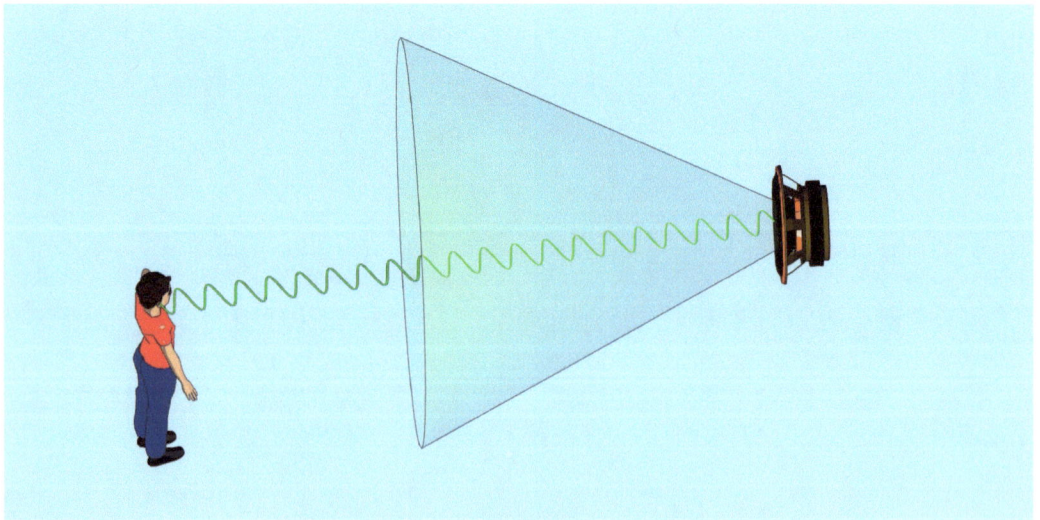

Illustration 47: What Tim really hears

Now, suppose that Tim stands near a brick wall. Sound bounces nicely off a brick wall, and there in an additional path from the speaker to Tim's ear:

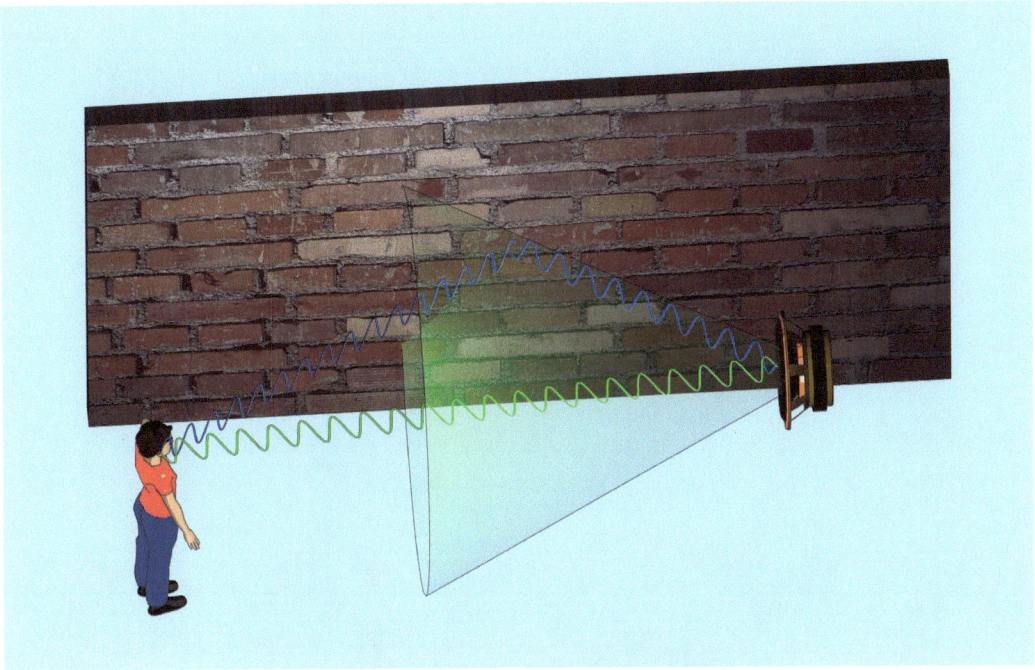

Illustration 48: Two paths for a wave

If you ask Tim to close his eyes and tell you what direction the sound is coming from, he won't be able to accurately tell you because it's coming from two directions. The volume of the sound will change as he walks around. Interestingly, as he walks around, it is relatively loud in some places and relatively quiet in other places. If the crests from the two paths hit his ear together, it will be loud. But if the crests from one path hit his ear with the trough of the other crest, they will cancel and the sound will be quieter in the same way that there were calm areas in the water on the ripple tank during the ripple tank interference experiments. He can actually walk around and find places where it's loud and places where it's quiet!

Furthermore, there's another path that you have to worry about – the path from the sound bouncing off the ground. This gives you an additional bounce. So suppose that you set up the dual source experiment by setting up speakers and don't worry about the wall or the ground as shown in Illustration 49.

Look at all those paths! There is a direct path from the right speaker in red and an additional red path bouncing off the ground. From the left speaker, there's the direct path in green, the path bouncing off the floor in green, and the path in blue bouncing off the wall. There's no hope of Tim being able to find the peaks and nulls from the direct paths because the additional paths are also interfering with them.

47

Illustration 49: Multi-path

For this reason, you have to make sure there are no walls or fences or anything else around for the sound to bounce off. You can do this by going out to a large field. To take care of the ground bounce paths, you can point the speakers up so that the ground is not within the cone pattern of the speaker. This is the arrangement we will be using for our experiments.

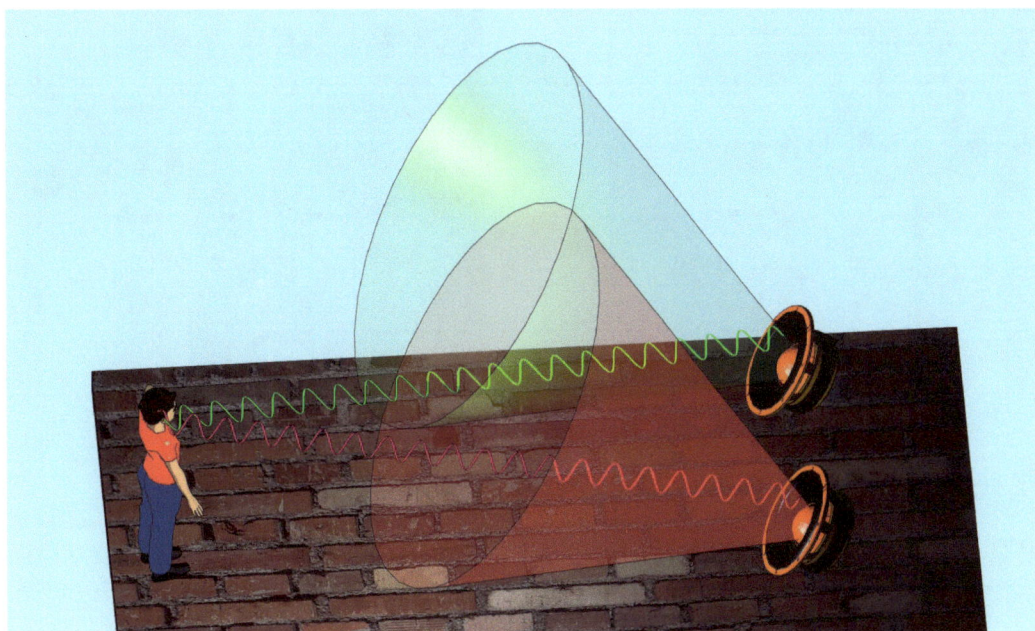

Illustration 50: Direct paths only

All the waves are missing the ground and there are no walls around to bounce off. Now Tim hears only the interference between the direct paths from each speaker.

What we will do is walk along paths parallel to a line between the two speakers. Where the waves constructively interfere, the sound will be loud. Where the waves destructively interfere, it will be quieter. It will not be completely quiet in these nulls, though, because the sound falls off as we get further from the speaker so when one speaker is close, it will be heard louder and not completely canceled by the quieter out of phase speaker.

Further, unless you are in an infinitely large field, there will be some things around and there will be *some* reflections coming back which will add other interferences. So rather than looking for sound/nosound, look for abrupt changes in volume. As you walk back and forth, drop a penny when the sound is at a minimum. Illustrations 51and 52 are pictures of Sam listening for nulls and then putting a penny down at one of the nulls. Note the curiosity of our dog. He was very interested in this activity. This surprised me – I did not think he would like the sound. As was pointed out by our neighbor, it was quite annoying to us humans.

Illustration 51: Listening for nulls

Illustration 52: Marking nulls with pennies

Notice he blocks the sound from his other ear to make it easier to find the nulls. This makes a big difference.

After doing this with a couple dollars worth of pennies, we had a map of the nulls. Because the pennies did not photograph well, we put large poker chips over the pennies for the purpose of this photograph in Illustration 53:

Illustration 53: Double source audio results with poker chips

To help see the nulls, I've drawn best fit lines in Illustration 54

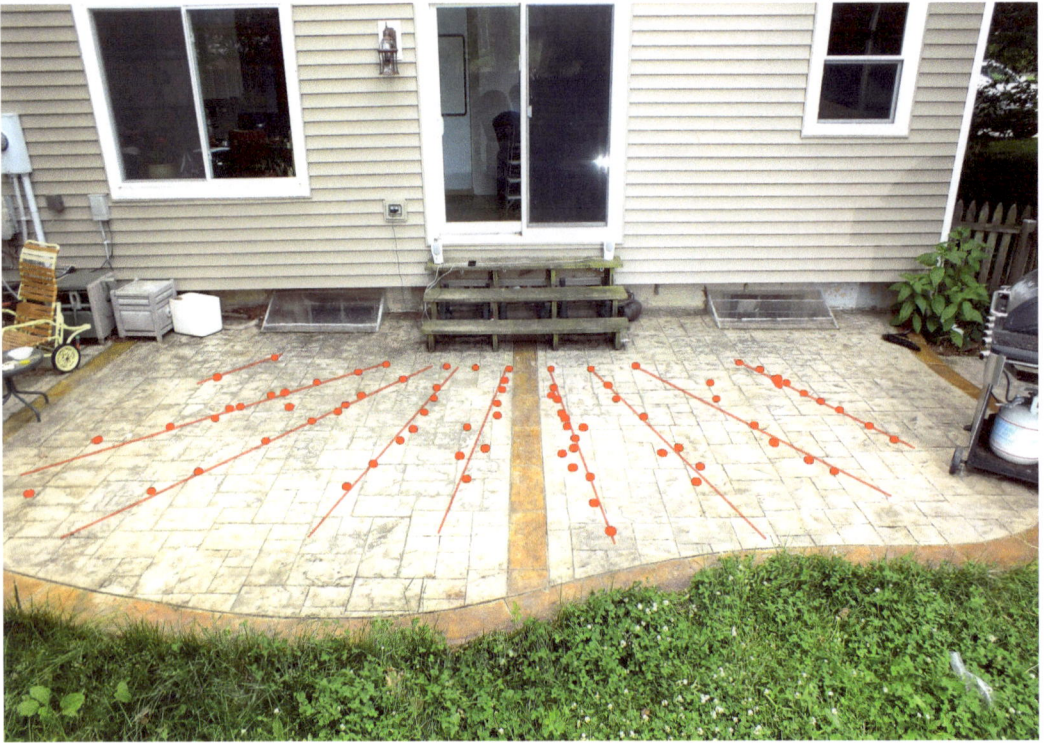

Illustration 54: Best fits lines for the nulls

Here's what it looks like if you hang out of a window in the second floor without and with the lines:

Illustration 56: Sound null results from above

Illustration 55: Sound null result lines

Now, compare that with an overhead pictures of pages 10-11 of *The Cat in the Box* and the ripple tank interference pattern:

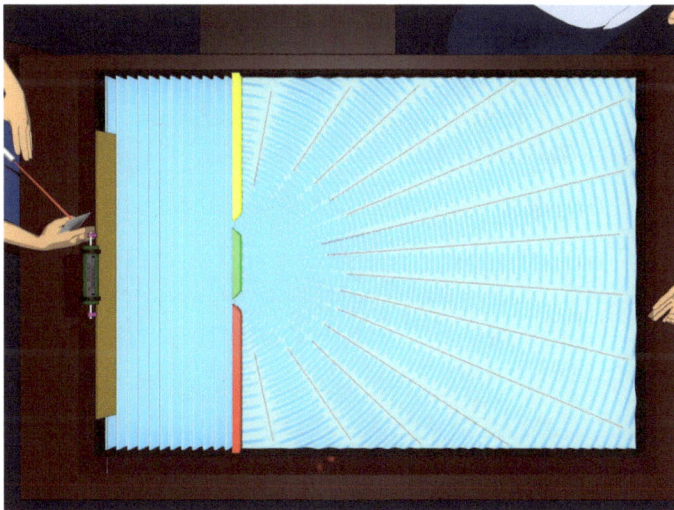

Illustration 57: Ripple tank calculated results from above

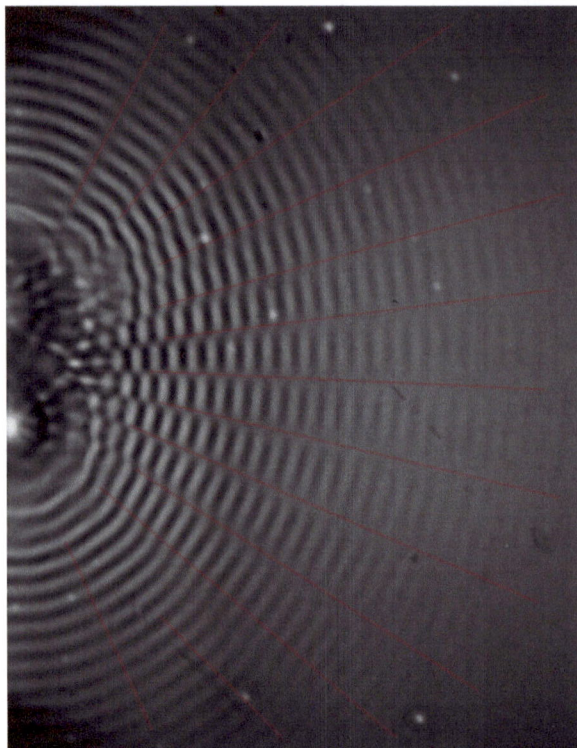

Illustration 58: Ripple tank results from above

The pennies are located where the distance between the speakers differs by half a wave (six inches), three halves a wave (18 inches), five halves a wave (30 inches), etc. The speed of sound depends on the temperature of the air, so the wavelength might not be exactly twelve inches. Knowing that the wave is approximately one foot, measure the distance from some of the pennies to each speaker and it and see how close the difference is from 6, 18, or 30 inches.

Experiment 6 – Double Slit Experiment with Light

This experiment is so common, I want you to find it in a book or on the Internet. You should definitely do this experiment though. The way most people do it is to cover a microscope slide with smoke from a candle and make slits in the smoke covering using two razor blades. A laser is then aimed through the slits onto a wall where you can see the interference pattern. From this, we conclude that light can indeed be described as a wave.

Usually, when one does the double slit experiment using light, you are seeing the results which are the equivalent of looking at the light at the rightmost side of the tanks, and you only see light and dark spots. The light spots are where the waves interfere constructively – where the ripple tank has active waves. The dark spots correspond to the regions of destructive interference and cancellation where the water is calm. You can see these spots in Illustration 35 (Cat pages 10-11) as Tim does the experiment.

It would appear that the light is not going through one of the slits, but going through both slits. One explanation for this is that we are shooting many photons and some of them go through one slit and some go through the other slit. The photons interfere with each other as waves, but at least there is some refuge in the thought that each individual photon goes through a single slit.

If this is the case, then if you shot the photons through one at a time, say one every second, there would be no other photons to interfere with and you would get the interference free pattern like the one with the rice. But when you do this experiment, you get the interference laden pattern! One interpretation is that the photon goes through both slits and interferes with itself. We shall look at this more later, along with other interpretations how a photon particle can be described with wave equations.

This series of double slit experiments was designed to determine whether something can be described a wave. It demonstrated that water waves, sound, and light are describable as waves, but sand is not. So, as you can see, there are experiments which clearly demonstrate that light has wave-like properties and other experiments which clearly demonstrate that light has particle like properties.

Summary of waves and particles

- Waves are a disturbance in some medium. The wave travels but the medium does not.
- Waves spread out as they go through small openings and bend around small objects in their path (diffraction)
- Waves add together when more than one of them travel though the same place. The sum of multiple waves can be interesting and very useful. (interference)
- Particles come in clumps and you can count the number of them. (countable)
- Particles do not diffract or interfere

Here is a chart summarizing what we've talked about with waves and particles:

	Diffraction	Interference	Quantized	Countable	Particle	Wave
Water	Yes	Yes	No	No	No	Yes
Rice	No	No	Yes	Yes	Yes	No
Sound	Yes	Yes	No	No	No	Yes
Light	Yes	Yes	Yes	Yes	Yes	Yes

Table 1: Waves and Particle test results

This chart sometimes stymies people because it shows that light has both particle and wave characteristics. You can do a number of experiments that show light "behaving" as a particle, and you can do a number of (different) experiments that show light "behaving" as a wave. We call this wave-particle duality.

Armed with this knowledge, we now return to the slit experiments.

Experiment 8 – Interference with electrons

A commonly question asked is whether photons are the only particles that have been shown to exhibit the wave particle duality. The answer is no. It has been shown experimentally with electrons, neutrons, protons, and other very small particles. It has also been shown with some larger things including a 60 atom buckyball made from carbon atoms. These are more difficult to do difficult to do, though and usually require a large facility.

Because it's not practical for most of us to do the electron slit experiments, we will describe the results. The electron version is important because we can add a few twists that can be done with electrons that can't be done with light.

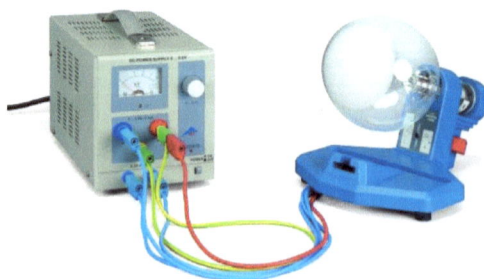

Illustration 59: An electron double slit experiment you could do at home

Actually, you can repeat the double slit experiment using electrons in your kitchen, but it takes a few thousand dollars of equipment. See, for example, http://www.telatomic.com/tubes/diffraction_tube.html.

A single photon and uncertainty

It is tempting to explain the results of the double slit experiment by saying that there are a large number of photons going through the slits and they are "simply" interfering with each other. One can apply the math of statistics to the problem and explain that half of the photons are going through one slit, the other half are going through the other slit. Because the photons also have wave

Illustration 60: The Cat pages 12-13

properties, they interfere with each other just like the many molecules of water arranged in waves going through the slits interfered with each other.

If you send rice through the cardboard double slit one grain at a time, there seems to be no doubt that you will get two piles of rice and no interference pattern. You can do this experiment, and you get just that – no interference pattern.

If you send photons through the double slit one at at time, however, the results are different. This experiment is harder to do because it's hard to send them through one at at time and it takes a looooooooooooooong time, but it has been done many times by many different people. Each one got an interference pattern. This is not like the rice results at all!

It's not just light that exhibits wave/particle duality. The experiment has been repeated for a large variety of things including electrons, protons, and buckyballs. Everything exhibits wave particle duality, but it can only be shown experimentally for very small objects. We'll go through the math for this in the *Wave effects of everyday objects* section. Cutting to the chase, the reason is that the larger the object, the smaller the interference pattern and for objects of everyday size, the interference is too small to be seen. A moving golf ball, for example, has a wavelength of only about 10^{-34} meters, which is smaller than an atom so we cannot see it's wave characters experimentally. But it is believed that *everything* exhibits wave particle duality.

The series logo is over the table in place of the light bulb.

It would be nice to know which slit each photon goes through before hitting the screen to make the interference pattern. This is not possible, though, because once a photon is detected, the photon is gone. So if we try to detect the photon as it goes through the slits, the photon is destroyed and does not make it to the screen. This is not the case with electrons - you can shine light at the path of an electron. The light is reflected off the electron and you see where the electron is. The electron does not get destroyed, and still hits the screen. So we can look at an individual electron, see which slit it goes through, and see where it later hits the screen.

Illustration 61 shows an arrangement to see which slit the electron goes through. If they use a beam of electrons, they would send trillions of trillions of electrons through the both slits at one time, and we would simply see lots of light coming from both slits. For this reason, they do the experiment one electron at a time. Doing it this way requires many hours of patient observation.

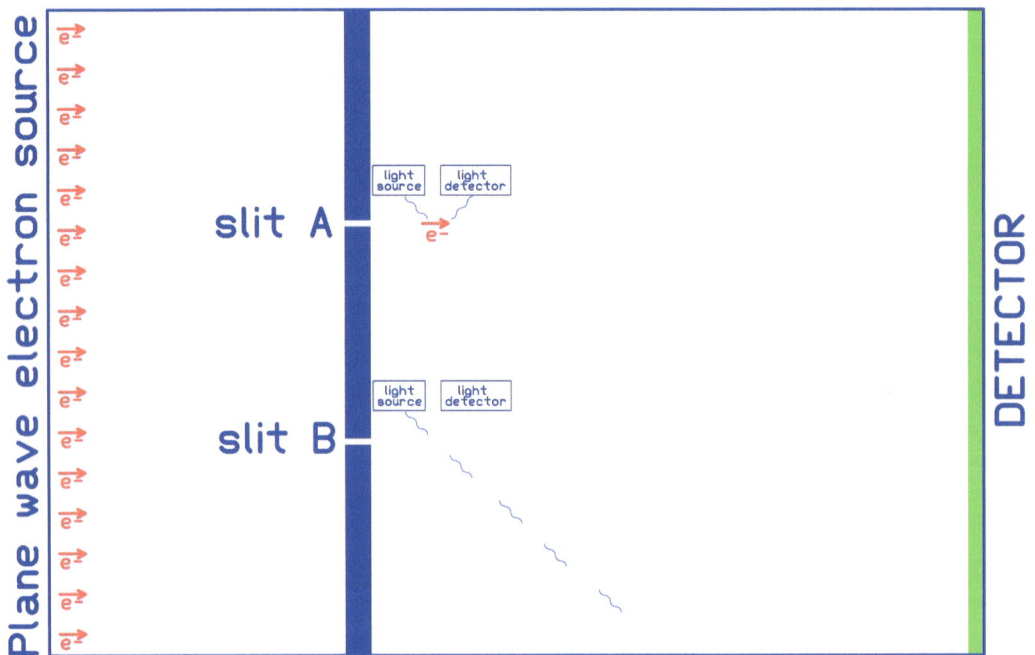

Illustration 61: Arrangement to measure which slit an electron went through

As a control, they begin by repeating classic double slot experiment with electrons instead of photons – sending lots of electrons through the slits and observing an interference pattern. The next step is to modify the experiment and send one electron through at at time so that they know the electrons are not interfering with each other – if an interference pattern shows up, it is because each electron is traveling as a wave, going through both slits, and interfering with itself.

The next step is to shine light (let's say blue light) on the slits and observe which slit an electron is going through (or, if they see light from both slits perhaps it is going through both). When they do this experiment, they find that each time an electron goes through, they see it at one and only one of the slits. But something else happens: the interference pattern goes away! When they do the experiment in the dark, they get an interference pattern. If they turn the blue lights on, they get no interference pattern!

So now they go back to the beam of electrons (as opposed to one at a time) and repeat that one with the lights on. Same thing: If they do the experiment in the dark, they get an interference pattern. If they turn the lights on, they get no interference pattern.

What's going on? The photons (particles) are colliding with the electrons (also particles) and disturbing their path. Imagine you are playing pool and take a carefully aimed shot to hit the solid green six ball into the corner pocket. As the ball is traveling toward the pocket, someone bombards the six ball with golf balls. The path of the six ball is

Illustration 62: Measuring something disturbs it

disturbed by the golf balls and misses the pocket. Once you interact with it, you force it to a mode which is best modeled as a particle. This is what the light is doing to the electrons which is why the interference pattern disappears.

No problem. We'll just turn down the light so that not so many photons (golf balls) are hitting the electron (six ball). When you do this, you find that even one photon hitting the electron disturbs its path enough to destroy the interference pattern (miss the pocket).

What if we use something with less energy than the blue photons so it doesn't disturb it as much? Recall from the section *What is light* when Tim got tired shaking the rope quickly that going to a longer wavelength light gives you lower energy photons. Imagine that instead of using a golf ball, you use something lighter. But to get lighter (less energy), it has to get bigger (longer wavelength). Idea: we could use red photons rather than blue ones. That way we wouldn't disturb the electrons as much. So suppose we use a balloon in place of the golf balls. Now the balloon will bounce off the billiard ball without disturbing it as much and we get an interference pattern. Unfortunately, the balloon (longer wavelength red photon) is so much larger than the space between the slits, you can't tell which pocket the six ball (electron) went in. Similarly, if the wavelength of light is sufficiently large enough to not disturb the interference pattern, it's too large to tell which slit the proton went through.

This is interesting - if we're going to use photons to see where the electron is, we have a problem. The shorter the wavelength light we use, the more accurately we can measure the position of the electron, but the more we disturb its path. The longer the wavelength light we use, the less we disturb the electron, but the less accurately we are able to determine the location of the electron.

So if we're going to use light to measure the path of the electron, we have to make a trade-off between measuring its position and disturbing its path. This is a fundamental limitation of measuring with light.

The Heisenberg uncertainty principle

It turns out that this limitation isn't just with light - it's with *any* measurement technique. Any attempt to measure which slit the electron (or any particle) goes through disturbs the particle sufficiently to wreck the interference pattern. The more accurately you measure the position, the less accurately you can know what the momentum was before the position measurement. This limitation was proposed by Werner Heisenberg in 1927 and is now known as the Heisenberg uncertainty principle. Many people, including Einstein, fought this bitterly, searching for measurement techniques that did not have this limitation, but to date no one has ever found a way around the principle.

Can you cite the Heisenberg uncertainty principle to get out of a ticket?

That is an interesting question.

> Officer: Do you know how fast you were going?
> Patron: No officer, I do not.
> Officer: Well, I measured your speed. You were going over 40 miles per

hour in a 25 mile per hour school zone.

Patron: How much over?

Officer: You were going exactly 42 miles per hour, young man!"

Patron: But officer, according to the Heisenberg uncertainty principle if you know exactly how fast I was going, you can't tell me where I was! So you can't know I was in a school zone.

Well, good luck with that.

Big uncertainty comes in small packages

The reason this ticket-avoiding argument won't work is that when things get big, like your car, the uncertainty in the position is too small to matter. Later in the *Wave effects of everyday objects* and *The effect of large sample statistics on Randomness* sections, we'll run through some of the math and see why this strategy won't work.

The difference between what happens in our macro world which is ruled by large sample statistics and the nano world which is ruled by quantum rules is an important concept. The uncertainty in something very very small can be significant. This uncertainty can be employed and is becoming useful in modern electronics and photonics. These principles are used in the design of quantum dots and tunnel diodes. But when large objects are being observed, the uncertainty is too small to be significant and things behave in the way we see in everyday life.

Let me summarize the results from the experiments we have so far.

1. Whenever we detect a particle, it interacts with us as a particle, coming in definite clumps: we never find half an electron or half a photon.
2. When it has not yet been detected or interacted with, its travel can be described mathematically as a wave: it diffracts around small objects and exhibits interference patterns. The smaller the particle is, the larger the interference pattern is and the easier it is to observe its wave characteristics.
3. Any attempt at observing the particle disturbs it. The act of measuring the particle changes the state of the particle: if you measure its position, you disturb its momentum. If you measure its momentum, you disturb its position. The more accurately you measure the position, the more you disturb its momentum.

Light cones and wavelets

The cat is playing with the light cone, under the table. This is the series logo and has two parts, a light cone and a Gabor Wavelet, representing the two cornerstones of modern physics.

A light cone is a tool used in relativity and a wavelet is a tool used in particle physics.

Our homework was finished, but we didn't feel done. We wanted more science because science is fun.

Tell us a story dear Dad! Tell us a story oh please. We want to have fun. That's all we need.

14   15

Illustration 63: The Cat pages 14-15

The cat is playing with the series logo under the table.

Light Cone

Consider living in Flatland, a two dimensional world like living on the surface of a piece of paper. You can travel north, south, east, and west but not up and down in Flatland[5]. I'm going to try to give you information about who won a basketball game between the University of Louisville Cardinals and the University of Kentucky Wildcats. Just as the game is over, I'm going to flash a light. If the Cardinals win, I'll flash red light and if the Wildcats win I'll flash a blue light. We then ask ourselves a simple question: Where and when can you know who won the game by watching my light?

First, we consider units. Are we going to measure distance in inches, meters, furlongs, or … ? We'll choose the interesting unit of light seconds, or the distance light travels in one second (186,000 miles or 300,000,000 meters). It looks instantaneous – when you turn on a light switch, the light seems to appear at your eyes immediately. But it doesn't – there is a delay but it is so short you can't

---

5   *Flatland* by Edwin Abbott ISBN 1623750318 is a lovely little book written in the 1800's about living in a two dimensional land. *Flatland* is the story of A. Square as he journeys outside his two dimensional universe through new dimensions. He experiences Spaceland (a universe with three dimensions), Lineland (a universe of one dimension), and Pointland (a universe no dimensions). Although some people – especially those who have read only the beginning of the book – consider the book to be sexist, it is actually a satire *against* the sexism of Victorian times. Aside from the political statements, *Flatland* is a very thought provoking book which teaches one to think in terms of non three dimensional universes and may of great assistance later in life when a child starts thinking about modern physics theories such as string theory.

easily observe it. Consider sound. Sound also appears to be instantaneous: when you clap, you hear the clap immediately. But it is easy to observe that sound takes time to travel: Go out to a field with a friend. Walk 100 paces away from your friend so that you're about 100 feet apart. Take two large disks that are easy to see 100 feet away (cymbals are ideal if you have some, if not two metal garbage can lids will work, or two large pot tops – anything that you can see from 100 feet away and make a loud noise when you hit them). Have your friend bang them together once while you watch. You will see them hit a tenth of a second before you hear them hit. Why? Because sound travels at about 1000 feet per second. To go 100 feet, it takes about a tenth of a second which you can notice if you observe carefully. If you can manage to get 1000 feet away (a little more than three football fields), the delay would be about one second. As you can see, it's not too difficult to measure the speed of sound.

Another great demonstration that the speed of sound is measurable is the delay between when you see lightning and when you hear thunder. Although they happen at the same time, there is a noticeable delay in the sound. How much? Sound travels at about a thousand feet per second and there are 5280 feet per mile. If we approximate this with 5000 feet per mile, it takes five seconds to go one mile. Since it takes five seconds to go one mile, if you notice three seconds between the lightning and the thunder, you are three fifths of a mile away.

And now a word from our light sponsor

With light, the experiment is not as easy because light travels faster. Much faster. About a million times as fast. Light travels so fast, the delay from 1000 feet is only about a millionth of a second, so it isn't enough to notice. To get a tenth of a second delay, you'd have to walk about a hundred million steps from your friend,

who would be too far a way to see[6]. To get a one second delay with light, you'd have to be about thousand million feet away. But there happens to be a convenient rock about 1200 million feet away – the moon. If we shine a laser at the moon, it takes just under three seconds for the light to travel to the moon, bounce off the moon's surface, and travel back. The Russians were the first to do this in 1962 by shining a laser on the surface of the moon and timing how long before they could see the spot using telescopes. The American Apollo astronauts left a retro-reflector (a special mirror arrangement that reflects light back to the direction from which it came) on the moon, making the measurement much easier to make. For more details, check out
http://www.lpi.usra.edu/lunar/missions/apollo/apollo_15/experiments/lrr/ and
http://en.wikipedia.org/wiki/Lunar_Laser_Ranging_experiment

And now, back to the game

Now that we understand that it takes light time to travel, let's see what happens when the basketball game ends and I flash my light.

Consider that the game has just ended. Louisville won, so I take my red light and turn it on and off quickly. One second later, the light has traveled one light second in all directions. If I did that on earth, the sound would travel out in all directions and the light's wavefront would be a sphere. Since we are on Flatland, the wavefront forms a circle.

Illustration 64: Light spreading out in Flatland for one second

Let's talk about what this means. Let's say that the light flashes on and off for a very short time at noon, which we shall consider to be t=0. The light travels outward in a ring. After one second, the time is t=1 and the light ring is as shown in Illustration 64. If you are inside the ring, you have already seen the flash. If you are outside the ring, you have not seen the flash yet. If you are right on the ring, you are a distance of one light second from the light and you are seeing the flash right now.

Illustration 65: Light spreading out in Flatland for two seconds

After an additional second, the light has traveled twice as far, so the circle's diameter is twice as big. After three seconds it has grown further, and after four seconds it is four times as big as it was after one second.

---

6   How far are we talking about? You'd have to walk almost all the way around the earth's equator!

After a third and fourth seconds,

Illustration 66: Light spreading out in Flatland for three seconds

Illustration 67: Light spreading out in Flatland for four seconds

Let's put these all on the same graph. The light hits that first circle at t=1s, the second circle at t=2s, the third circle at t=3s, and the fourth circle at t=4s. So if you are on that outer circle, you will see the light flash four seconds after it actually flashes.

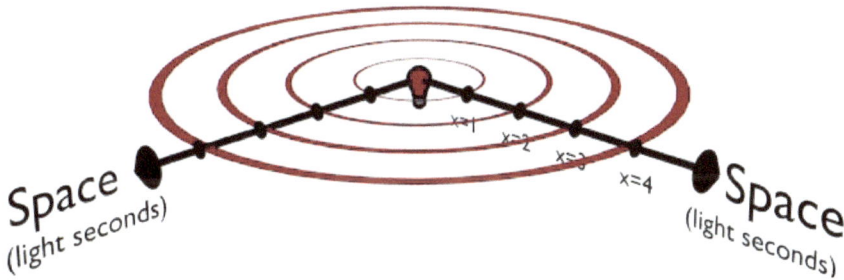

Illustration 68: Circles drawn where light is at the first four seconds

Now, consider that I take these four drawings and stack them one over the other so that I have made a 3D graph, where going up one tick mark represents waiting one second:

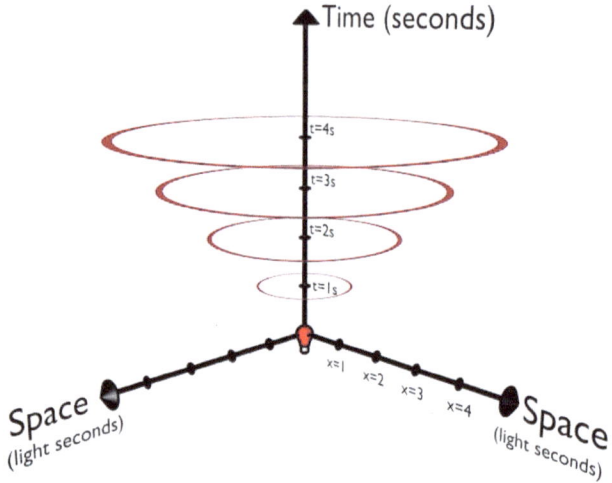

Illustration 69: Stacking the circles vertically with time

If we measure twice as often, we get the following drawing:

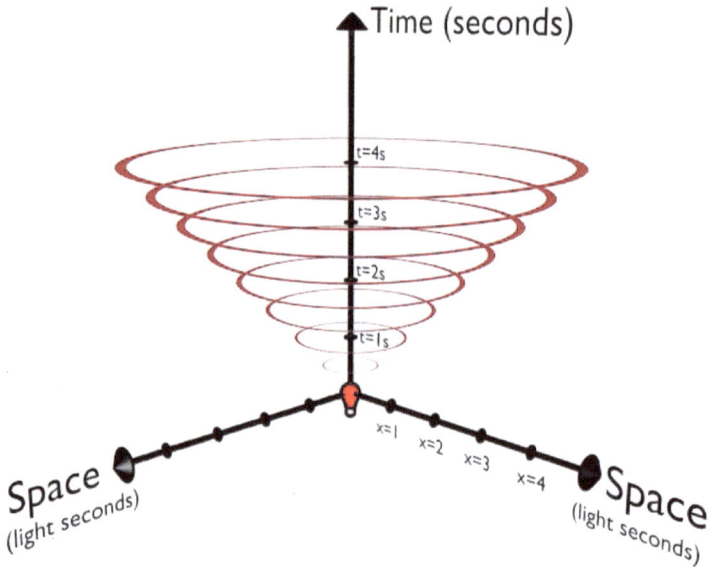

Illustration 70: Measuring more often

If we measure continuously, we'll get a nice continuous surface:

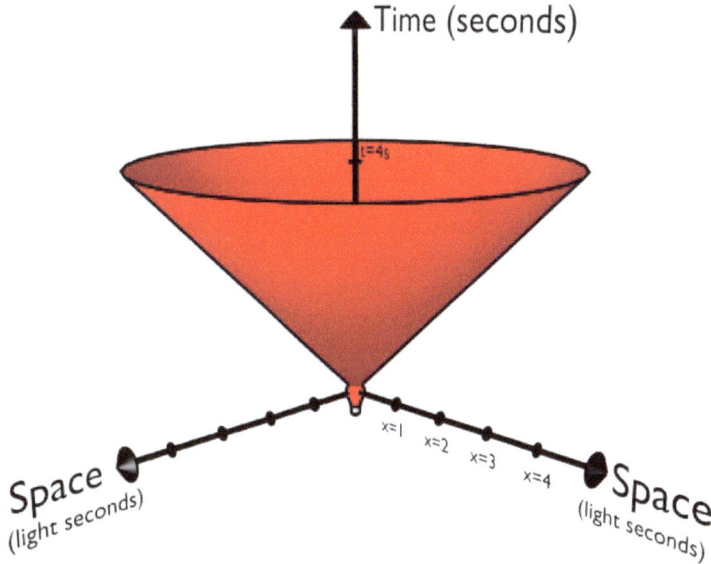

Illustration 71: Measuring continuously gives us a full light cone

This is called the future light cone and it shows when and where someone could see my light after I turn it on. Inside the cone someone can know who won the game. Remember, this is a time-space diagram, so any particular point marks a time *and* a location in Flatland. So if I am located inside the cone, I have already seen the light but if I am located outside the cone, I have not yet seen the light because it hasn't gotten to me yet. If I am on the cone's surface, I am seeing the light right now.

Now consider a similar question: Suppose I want to know who won the Boston Celtics – Los Angeles Lakers game. I will ask someone at the game to flash a green light if the Celtics win and a purple light if the Lakers win. The question is: From where and when could someone have sent me a light indication of who won the game if I'm receiving it right now? Using the same technique of drawing circles, I find that the past light cone looks like this:

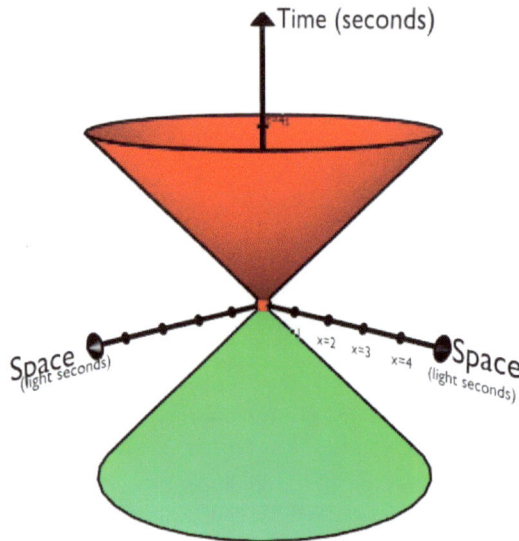

*Illustration 72: Adding the past, we get the past
and future light cones*

It looks like the Celtics won.

Our series logo contains the past and future light cones. This simple tool turns out to be very useful when looking at space time questions, especially in light of relativity (pun intended!). When you learn about relativity, the light cone is indispensable.

Wavelet

A wave is a moving disturbance in a material. In general, when we talk about waves we are talking about a disturbance that has been going on for a long time, practically forever. A wavelet is a small section of a wave, or a short-lived wave. If you kick your feet at the edge of a pool for a long time, you will create a wave in the water. If you toss a small rock into the water, you will create a wavelet.

Things in our universe are made of waves. Everything has a wave function including photons, electrons, and your cat. The wavelet in the logo represents wave functions. Wave functions are used for a variety of purposes in modern physics including answering the seemingly simple question: what is the probability that something will be found in a particular place at particular time?

Suppose that someone tosses a ball to you in a completely dark room. You want to know when and where the ball is so you can be in the right place at the right time to catch it. You turn on a red light for a very short time so that you can see where the ball is right when the light is flashed. The longer you leave the light on, the more the image of the ball is smeared, so you flash it as quickly as you can.

When you flash the light, some of the light bounces off the ball into your eye and you can see where the ball is. There is a little bit of a delay due to the time it takes for light to travel the ball to your eye, but since the light is traveling so much faster than the ball is moving we can ignore that delay.

Because light is a wave, what hits your eye is short pulse of a wave, or a wavelet. The wavelet is spread out in space time, or *smeared*. We can't know the location of the ball any more accurately than the size of the wavelet. In order to get more accurate a location, we could use a shorter pulse of blue light which has a smaller wavelet.

The interaction between the light and ball disturbs the speed of the ball which is randomly changed with that interaction. The shorter the wavelength, though, the more it disturbs the speed of the ball. So if I measure the location of the ball more accurately, the less I will know about its speed. This is the basis of the Heisenberg Uncertainty Principle we spoke about earlier in the *A single photon and uncertainty* section: independent of how I measure things, the more I know about something's location, the less I can know about its momentum (speed times mass).

Our series logo contains a simple wavelet, perhaps the wavelet of a single particle. Of course the wave function for your cat is much more complicated. Wave functions are useful tools for particle physics.

Is relativity really real?

The book on the left side of the table is *Relativity,* a book Einstein wrote in an attempt to explain relativity to the layman. This book, and many others like it, attempt to put relativity into terms that non physicists can understand. Some attempt to do this by minimizing the math and using analogies to try to get the concept across; others fully utilize the math.

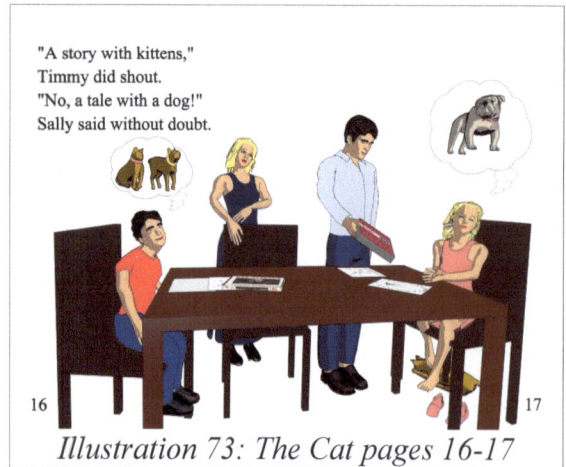

"A story with kittens," Timmy did shout. "No, a tale with a dog!" Sally said without doubt.

Illustration 73: The Cat pages 16-17

But it's not the mathematics that makes the concept difficult. It's the counterintuitive nature of relativity with respect to humans living on earth that makes the concepts foreign and unbelievable. There are no useful analogies because we don't experience anything like relativity in everyday life.

Because many people find that it goes against their common sense, they do not believe in relativity. My brother is one of these people.

I often joke that my brother disproved Einstein's first claim about relativity, the claim right there on the cover of the book: "A clear explanation that anyone can understand."

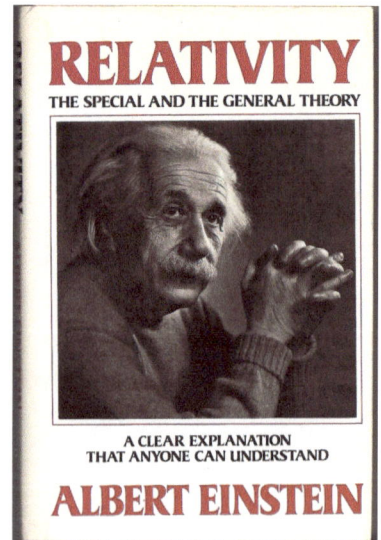

Illustration 74: Einstein's explanation of Relativity for all audiences

How is it that a well-educated man from such a fine family can completely miss out on relativity? Relativity runs counter to our everyday experience. Our everyday experience – and all the math we learned the first eighteen years of our life – tells us that if I'm going 50 miles per hour due north and you are going 50 miles per hour due south and we pass each other at noon, an hour later we will be agree we are 100 miles from each other. Relativity says this isn't true, and this is a hard pill to swallow[7]. This is why we want to expose people to relativity at a relatively young age. This early experience will allow them to develop an intuition compatible with relativity.

 The logo is small on this page and is located on the dog's collar.

---

7   See the example in the *Introduction* at the beginning of this book.

The Feynman lectures on physics

The book Dad picked up is the Feynman Lectures on Physics. In 1961, Caltech asked Dr. Richard Feynman to revamp how physics was being taught.

Along with Dr. Leighton and Dr. Sands, Dr. Feynman re-wrote Caltech's introductory physics

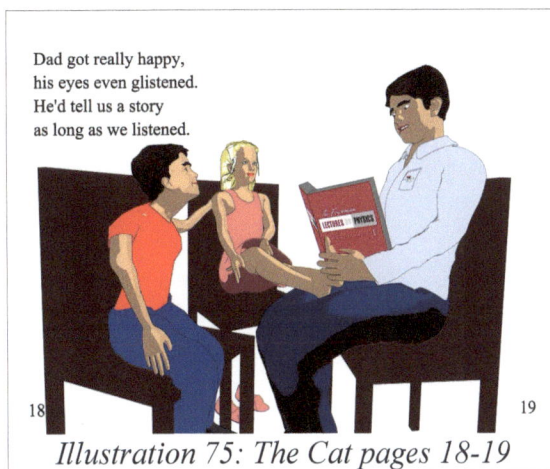

Illustration 75: The Cat pages 18-19

course. The notes from the course were put together into a 3 volume set (ISBN 0201021153). In general, this is not a light read but a thoroughly enjoyable, elegant and thorough course in physics and an enjoyable read by someone already familiar with physics and a rudimentary understanding of the subject.

The series has enjoyed amazing longevity with a plethora of publications over the last 50 years including a commemorative set, Six Easy Pieces (ISBN 0465025277) which is a set of the easier to understand chapters, Six Not So Easy Pieces (ISBN 0465025269), and the Definitive and Extended Edition (ISBN 0805390456) which includes a fourth book, Feynman's Tips on Physics.

If you'd like to Listen to Feyman deliver the lectures, they did record the lectures in 1961 and 1962 on a reel to reel recorder. Caltech kept the recordings all these years and they are available on CD (for example ISBN 0738209244) and are also widely available on the Internet. I highly recommend getting some of these – they are thoroughly enjoyable and educational. A good starting point is to listen to the Six Easy Pieces on audio (ISBN 0201483068). These, too, can also be found on the Internet.

The series logo is on the pocket of Dad's shirt.

The miracle year of 1905

1905 is known as Albert Einstein's Miracle year. In that year, he authored five major papers, each of which made a substantial contribution to physics:

1. For his Ph.D dissertation, he **determined the size of molecules** by carefully and cleverly measuring the viscosity of dissolved sugar solutions.

Illustration 76: The Cat pages 20-21

2. He described light as particles (photons) with a discrete amount of energy rather than a wave. This paved the road to a huge variety of discoveries including explanation of black body radiation, the field of quantum physics, and the interaction between light and electricity, known as **the photoelectric effect.** It is hard to overestimate the impact of this work and he was awarded the 1921 Nobel Prize in Physics for this contribution.

3. He **proved atoms existed.** In the 1850's Robert Brown observed through a microscope that small things move in water with no apparent source of energy. This is now known as Brownian Motion. Although the idea of atoms had been proposed by the ancient Greeks and Indians and others, in 1905 the existence of atoms was not well accepted and widely controversial. Einstein showed that if atoms existed, they would cause small things to move in exactly the way Brown observed. This provided strong support to atomic theory and ended the controversy.

4. He explained the curious result of the Michelson Morley experiment which inexplicably measured the speed of light to be a constant independent of your motion relative to the light source. This famous work is known as **the special theory of relativity.**

5. He unified mass and energy, culminating in **the famous equation E=MC^2**. This was the birth of many technologies including MRI machines, unification of mass and energy, and invention of nuclear energy.

Further information can be found in John Stachel's enjoyable *Einstein's Miraculous Year*, ISBN 0-691-12228-8. In this deeply technical book, Stachel presents the five papers along with some wonderful background and explanations.

The series logo (time cone only) is written on the blackboard.

More miracle years

How and why Einstein was so productive in one year is of much discussion. He was just out of school and in the mode of deep thinking and hard work. He had a job which wasn't demanding[8] which meant he had time to spend and could spend his intellectual capacity on personal projects such as developing relativity. At age 26, Einstein was well posed in 1905.

1905 was not the first miracle year. The Wikipedia article on Annus mirabilis (Latin phrase meaning *wonderful year*) lists a dozen or so miracle years. Some of the more noteworthy years relevant to our subject include 200BC, 1543 and 1666.

In 200, BC, Archimedes determined the ratio of the volume of a sphere to a cube just enclosing it. It's interesting that until recently we did not know Archimedes knew about this and thought this fact wasn't discovered until Newton's time, 1200 years later. In 2010 an ancient religious prayer book was purchased at a Sothby's auction by a wealthy enthusiast. He funded a detailed analysis of the book. The prayer book was made from recycled parchment, which came from a transcription of an Archimedes scroll. X-ray fluorescence was needed to decipher pages overlaid with gold leaf. Determining the formula to find the volume of a sphere requires calculus. It wasn't until 2010 that we knew Achimedes invented calculus over 2000 years ago.

In 1543, at age 28, Andreas Vesalius published *On the Fabric of the Human Body*, leading to the development of study of human anatomy and the practice of medicine. Nicolaus Copernicus, at age 70, published *On the Revolutions of the Heavenly Spheres*, which not only paved the way for the science of astronomy, but opened the minds of people to acceptance of the scientific method.

In 1666, at age 24, Isaac Newton described motion in his three famous laws of motion, developed a theory of gravity, developed a theory of light and colors, and re-invented calculus, taking it much further than Archimedes and into the mainstream where it enjoys everyday use in science, engineering, and other fields. Not a bad year!

---

8 Many biographers have noted that his job wasn't very demanding. My take on this: well, perhaps it wasn't demanding *to him* - he was a patent clerk which might have been quite intellectually challenging to me!

Busting quantum theory

Einstein's work in 1905, especially the work on the photoelectric effect which would win him the 1921 Nobel prize in Physics, was the beginning of quantum physics. In this paper, he noted that light came in discrete packets of energy.

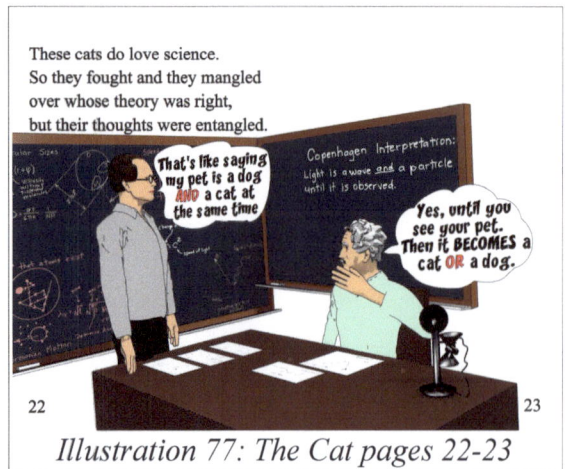

Illustration 77: The Cat pages 22-23

Quantum theory wasn't invented one afternoon by a single person. The founders of Quantum Theory listed in Wikipedia article *Quantum Theory* are Max Plank, Niels Bohr, Werner Heisenberg, Louis de Broglie, Arthur Compton, Albert Einstein, Erwin Schrödinger, Max Born, Paul Dirac, Enrico Fermi, Wolfgang Pauli, Max von Laue, Freeman Dyson, David Hilbert, Wilhelm Wien, Satyendra Nath Bose, and Arnold Sommerfeld. The majority of the theory was developed between 1905 and 1935, and it took a lot of discussion to formulate this theory. Einstein was not exactly a willing partner in its development.

Einstein did not like many of the concepts of quantum physics, and spent much of the remainder of his life challenging the very theory which he had invented. Quantum physics makes a lot of predictions which are counterintuitive. One of its predictions is that the more accurately you measure the speed of a particle, the less you can know about its position. This law was discovered by Werner Heisenberg in 1925 and is known as the Heisenberg uncertainty principle. See the section *A single photon and uncertainty* for more information. Einstein did not accept the Heisenberg uncertainty principle as foundational. He acknowledged that measurements of the day did indeed yield that result, but he argued that no one had yet invented the right measurement technique and that the theory simply wasn't yet complete. Surely, the measurements could be made.

In 1935 Einstein, Boris Podolsky, and Nathan Rosen published a paper *Can Quantum-Mechanical Description of Physical Reality Be Considered Complete?* - This famous paper is known as the EPR paper and established entanglement which Einstein hoped would deliver the death blow to the Quantum Theory. They set up a thought experiment which included a method to measure both the speed and position of a particle. This ingenious arrangement was meant to once and for all disprove the Heisenberg uncertainty principle.

The series logo is part of the telephone.

We start with an unstable atom which decays and radiates one electron (β radiation) and one positron (positron radiation). A positron is very similar to an electron except that it has positive charge rather than negative charge. All the other aspects of a positron are the same as the electron, for example they have the same mass. The decay created one positive charge and one negative charge, so charge is conserved as we expect (see the earlier section *Conservation of Charge*).

Electron positron emission

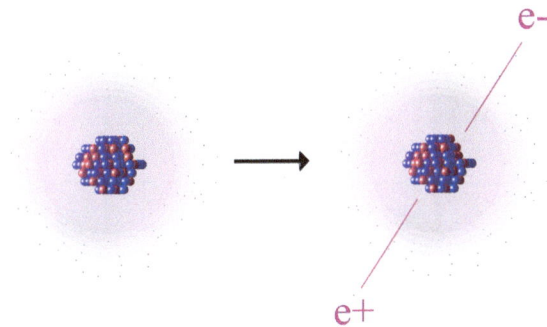

Illustration 78: An unstable atom creating an electron/positron pair

Charge is not the only thing that the universe conserves. Many things are conserved including the product of mass m and velocity v. This product mv turns out to be so convenient in calculations that we give it a name: mass times velocity is called momentum. So we say that, like charge, momentum is conserved. If we start out with a total momentum of 5, we end up with a total momentum of 5, regardless of what happens during the experiment.

Let us assume that the original unstable atom is not moving at the beginning of the experiment. The total momentum is therefore zero. At noon, suppose the atom decays and spits out an electron going to the left and a positron going to the right. After the decay, suppose the atom is still not moving. Since momentum is conserved, the speed of the electron must be exactly the same as the speed of the positron.

Applying the Heisenberg uncertainly principle to the electron, if you measure the location of the electron very accurately, you will randomly disturb its momentum and therefore can't go back and accurately determine the original momentum. Einstein, Podolsky, and Rosen proposed to measure the position of the electron and the momentum of the positron. Since the momentum of the positron is the same as the momentum of the electron, one would then know both the position and momentum of the electron. They yielded that measuring the position of the electron randomly changes its momentum. But unless measuring the position of the electron also randomly changes the momentum of the positron, this would mean a violation of the Heisenberg Uncertainly Principle.

Many people (including Einstein) had previously proposed methods to violate the Heisenberg uncertainly principle. But Heisenberg had always found something wrong with their method. For a while, no one could find anything wrong with the Einstein, Podolsky, and Rosen argument. It did seem that unless measuring the position of the electron randomly and instantaneously changes the momentum of the positron, one could violate the Heisenberg Uncertainly Principle.

Eventually Heisenberg had to concede that Einstein was right: unless measuring the position of the electron randomly changes the momentum of the positron, one could violate the Heisenberg uncertainly principle. Instead of conceding that the Heisenberg uncertainly principle was wrong, however, Heisenberg proposed that measuring the position of the electron *did indeed* instantly change the momentum of the positron. Einstein did not like this and sarcastically called it "spooky action at a distance." As ridiculous as this seemed, experiments show that this is the case. Whatever happens to the electron also happens to the positron. This is now known as quantum entanglement. Electrons and positrons are not the only particles that can be entangled. Scientists are routinely creating entangled photons and using them for some encryption methods which promise to provide more secure communications than any current methods. Protons been entangled, and so have buckyballs and even larger molecules. Some people are working on quantum teleportation using entanglement.

Einstein did not get what he wanted. Rather than the paper dealing a death blow to quantum theory, it established the field of quantum entanglement. Einstein unwillingly made a *lot* of contributions to quantum theory.

Can you believe this?

Here's what Feynman had to say:

"Things on a very small scale behave like nothing that you have any direct experience about. They do not behave like waves, they do not behave like particles, they do not behave like clouds, or billiard balls, or weights on springs, or like anything that you have ever seen." …

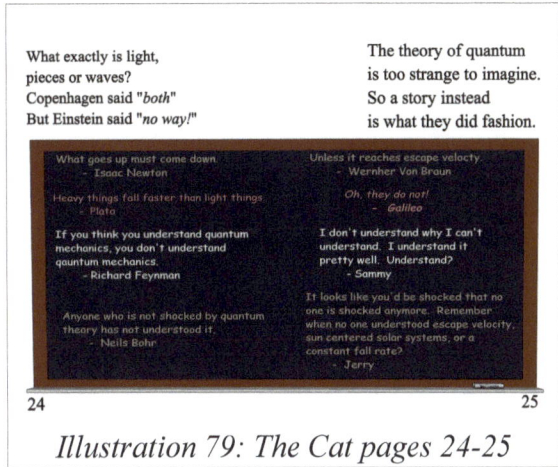

Illustration 79: The Cat pages 24-25

"Because atomic behavior is so unlike ordinary experience, it is very difficult to get used to, and it appears peculiar and mysterious to everyone – both to the novice and the experienced physicist. Even the experts do not understand it the way they would like to, and it is perfectly reasonable that they should not, because all of direct human experience and human intuition applies to large objects. We know how large objects will act, but things on a small scale just do not act that way."

...

"I have asked you to imagine these electric and magnetic fields. What do you do? Do you know how? How do I imagine the electric and magnetic field? What do I actually see? What are the demands of the scientific imagination? Is it any different from trying to imagine that the room is full of invisible angels? No, it is not like imagining invisible angels. It requires a much higher degree of imagination. ... Why? Because invisible angels are understandable. ... So you say, "Professor, please give me an approximate description of the electromagnetic waves, even though it may be slightly inaccurate, so that I too can see them as well as I can see almost-invisible angels. Then I will modify the picture to the necessary abstraction."

"I'm sorry I can't do that for you. I don't know how. I have no picture of this electromagnetic field that is in any sense accurate. ... So if you have some difficulty in making such a picture, you should not be worried that your difficulty is unusual."

The series logo is on the piece of chalk.

With all due respect to Feynman, and that is a *very* high level of respect, Sam and I feel strongly that this is an addressable issue.

In fact, Feynman's claim could even give us a blueprint for what we need to do to solve the very problem he is describing,

He mentions that light does not "behave" like anything that you have ever seen and that "Because atomic behavior is so unlike ordinary experience, it is very difficult to get used to, and it appears peculiar and mysterious to everyone."

What we intend to do, then, is to show children the characteristics of light and give them experience in the microscopic world. Give them stories that include the effects of relativity and quantum physics so that when it is time for them to take courses in modern physics, they have experience and do not find the behavior peculiar or mysterious.

If this seems far fetched, consider how children who grew up in the 21st century approach the Internet. Using the Internet requires a huge amount of knowledge and experience. Like anything else, when built up slowly over over one's life, it is quite manageable. But if dropped into learning it quickly without this experience, it is quite challenging.

No, no Grandpa. "Double click" doesn't mean click both buttons at once.

Illustration 80: Sigh. But it seems so easy!

Accepting quantum mechanics need be no harder than accepting that everything in a vacuum falls at the same rate. Accepting relativity need be no harder than accepting that the world is round (*spherical*, actually).

Schrödinger's Cat

By 1935, Einstein and Schrödinger had had enough of this "crazy" quantum theory that claimed not only that actions were random, but that actions really didn't happen until they were observed. Much argument over this ensued and Einstein's answer to all this was his famous quote "God doesn't roll dice with the universe!"

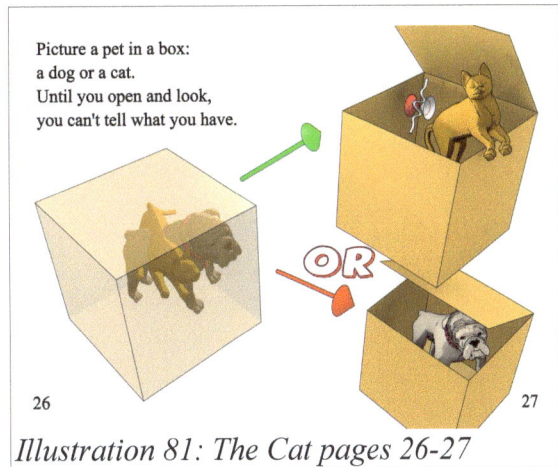

Picture a pet in a box:
a dog or a cat.
Until you open and look,
you can't tell what you have.

OR

26 27

Illustration 81: The Cat pages 26-27

There are different ways to think about and interpret the math behind quantum theory. In 1935, quantum theory of the day explained the results of the double slit experiment by saying that the photon went through both slits until someone observed it. When it is observed, it *becomes* having gone through either the top slit or the bottom slit. It doesn't decide at the time of observation – the path happened in the past. At the time of observation, the photon goes back in time so that the path *became* one of the slits. This concept is bizarre, but explains what happens in every experiment. In other words, the mathematics describing this concept correctly predict the results of the slit experiments. This math can be interpreted in many ways. The interpretation that both possibilities exist until the result is observed was developed mainly by Werner Heisenberg and Neils Bohr when they worked together in Copenhagen and is known as the Copenhagen interpretation. We will look at some other interpretations later.

When extended into the macroscopic world such as flipping a coin, the coin is both heads *and* tails until someone looks at it, at which time it *becomes* either heads or tails. From Schrödinger's perspective, this was utter nonsense and together with Einstein he came up with a thought experiment that everyone could relate to. This thought experiment became known as Schrödinger's cat.

Their original thought experiment is not really kid friendly, so we're taking the liberty to change it a bit. The original one is described in the lost pages section of this book.

We now describe in kid friendly terms the famous Schrödinger's cat thought experiment as depicted on pages 26 and 27 of *The Cat in the Box*.

The series logo is in the box with the cat.

As mentioned, there are different ways to think about and interpret the math behind quantum theory.

The Copenhagen interpretation of quantum physics

Consider a box where a cat or a dog has been randomly placed into the box. For example, a dog, cat, and empty box were in a room and the box was closed. Consider a Rube Goldberg machine that detects cosmic rays and opens one side of the box letting in one animal and closing the box. The door to the house is then opened and the other pet gets away undetected. The pet inside the box has been decided by a cosmic ray, which is considered to be a random quantum event.

The unknown pet is inside a closed box, and the Copenhagen interpretation of quantum theory says that until you look into the box (or smell it, or listen for purring, or x-ray it, or make an observation in *any* way), the single pet is both a dog and a cat at the same time. When you look into the box, the pet instantaneously becomes a dog or a cat but not both. But before the observation, the Copenhagen interpretation says the pet is both a dog and a cat. This has been experimentally verified many many times.

What does this mean? Suppose you develop a mathematical model for the pet being both a cat and a dog until it is observed and then the math models a dog or a cat after the observation. When you apply that mathematical model to predict the outcome of quantum experiments, the model will correctly predict the results of quantum experiments such as the double slit experiments.

Over the last 50 years, every experiment done agreed with the predictions from theory that the cosmic ray has both come and not come at the same time and that the pet is both a dog and a cat at the same time.

If this intrigues you, you will find John Wheeler's ingenious delayed choice experiments fascinating. We will discuss Wheeler's delayed choice experiments later in the lost pages section of this book.

This goes against our everyday experience at such a fundamental level that we have great difficulty accepting it. *Nevertheless...*

What *can* you know?

The concept on this page is that until one opens the box and makes an observation, one cannot know what's in the box. It isn't simply that you can't see whether it's a dog or a cat, but that it *isn't* a dog or a cat before the observation is made. It's not just that Schrödinger doesn't know, it's that he *can't* know because it isn't yet a dog or a cat. Remember, according to the Copenhagen interpretation, all

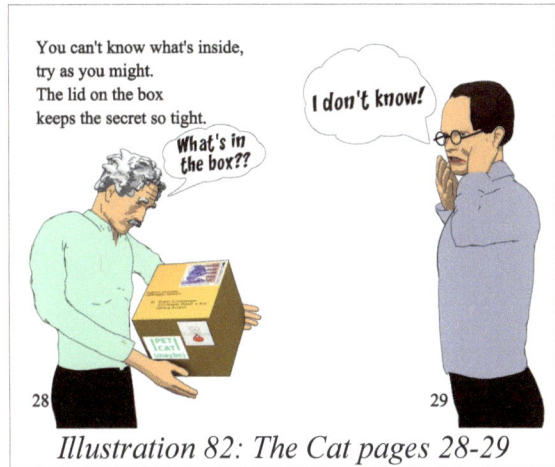

Illustration 82: The Cat pages 28-29

possibilities exist until it has interacted with something or been observed by someone.

The large stamp on the top of the box is a stamp with one of Einstein's most famous quotes, "Imagination is more important than knowledge." To become a success at modern physics, kids have to learn this in spite of what teachers tell them. And I'm saying this as one of those teachers.

This is very important to the concept of the *Mom, I wanna be a Scientist* series because imagination is most easily instilled at an early age.

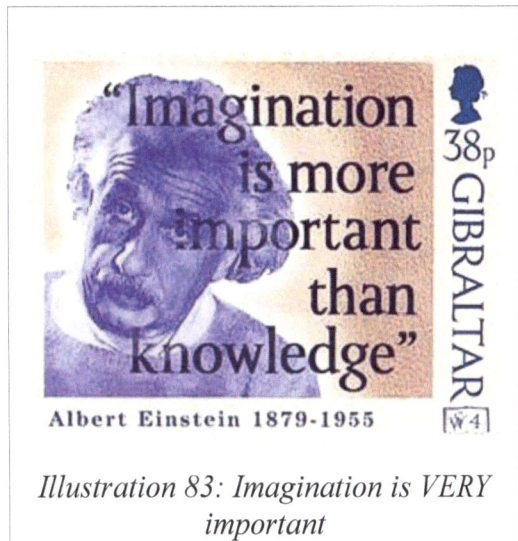

Illustration 83: Imagination is VERY important

The series logo is on a sticker on the outside of the box.

Statistical interpretation of quantum physics. Who knows?

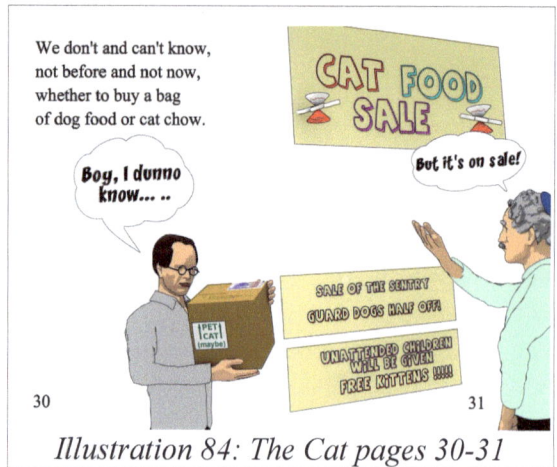

In the early and mid 20th century, they talked about the pet becoming a dog or cat when someone observed the pet. It was both until someone looked into the box and make an observation. This developed into a philosophical question of who counted as an observer. Did it become a cat or dog for everyone when Schrödinger's observed it, or was it still a cat and a dog for Einstein until he himself observed it?

Illustration 84: The Cat pages 30-31

And who counts as an observer? Does the pet count? Does a digital camera count? Does a film camera count before the film is developed? People used to spend a lot of time wrestling with these questions. We will talk about this more in the lost page section *Who counts as an observer?*

In recent years, most people have stopped thinking about it in this way. Today we don't worry about if the observer has to be sentient. Rather, we think about when the object in question interacts with something large. The object is observed when the quantum effect is amplified to the macro world and the numbers are large enough to be ruled by statistics. This is an important concept. The randomness happens in the micro world, but when you have 12 grams of carbon, 602,200,000,000,000,000,000,000 atoms, the randomness cancels out and it becomes practically deterministic. For example, if I'm flipping an honest coin twice you cannot predict the number of tails with much certainty. If you make a prediction and we do the experiment a few times, you will have a 50% average error. But if I'm flipping it a million times and you predict that there will be half a million heads within five percent, you will likely have a less than 0.00001% error. This you can count on and provides the basis for the deterministic nature we see in our macro world even though it comprises of random events. We'll go into more detail on this in the *The effect of large sample statistics on randomness* and *The statistics of coin flipping* sections.

The series logo is on the CAT FOOD SALE sign.

Many worlds interpretation of quantum physics

In 1957 Hugh Everett took a close look at the math of quantum physics and developed an interpretation different from the Copenhagen interpretation. When there was a choice such as dog versus cat, the Copenhagen interpretation says both possibilities exist until an observation is made and then instantaneously only one exists. Remember, the math predicts the results of the experiments and is therefore "deemed correct."

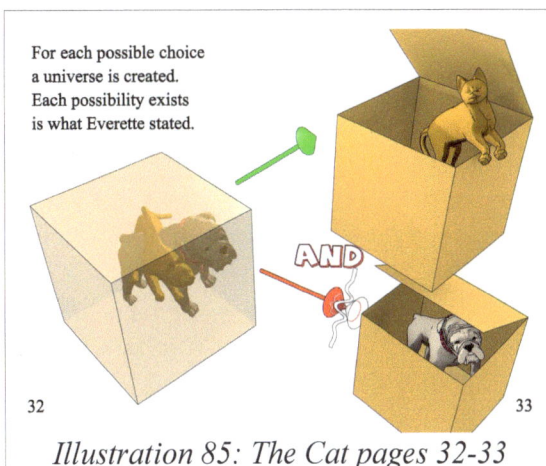

Illustration 85: The Cat pages 32-33

Hugh Everett observed that the same math could describe another phenomena: whenever more than one possibility exists, they are both true, but in separate universes. When both a dog and a cat could be in the box, the universe splits into two separate universes. No one in the "dog universe" can interact with anyone or anything in the "cat universe." Since it's the same math, this interpretation also predicts the results of the experiments and is also "deemed correct."

A few years later Bryce DeWitt coined the term "Many worlds interpretation" to describe Everett's interpretation of the wave particle mathematics.

There are many different interpretations which can be described using the the same math. The Copenhagen interpretation is best known because it was the first widely accepted (and widely debated!) interpretation and the appeal of the Schrödinger's cat analogy. The many worlds interpretation is well known because of its appeal. It's *fun* to think about.

Notice that Illustration 85 is nearly the same as the Copenhagen interpretation from pages 26 and 27. There is a difference though: the OR has become an AND. In the Copenhagen Interpretation, *one* of the possibilities become reality. In the many worlds interpretation, *all* of the the possibilities become reality.

There is no communication between the different worlds, however. In the case of the dog/cat possibilities, an observer who sees the dog will never see the cat. Many people have looked for ways to provide some communication between different worlds. So far no one has found a way.

The series logo is at the end of the red arrow.

83

Is good math a good reason?

This brings up an interesting question: Is "because the mathematics works" a good reason to have confidence in something? The answer is that it is certainly a good enough reason to make it a strong candidate deserving a closer look. The mathematics and especially the units can be used to rule something out if it does not match and make it a strong candidate if it does match.

Let's say you're taking a physics test and have to calculate how fast you need to travel to go 15 miles when you have 3 hours for the trip. Let's say that you cannot remember the formula for speed. You're pretty sure that it's either time divided by distance or distance divided by time. So you try both, treating the units just like numbers:

$$speed = \frac{3\ hours}{15\ miles} ? = 0.2\ \frac{hours}{mile}$$

$$speed = \frac{15\ miles}{3\ hours} ? = 5\ \frac{miles}{hour}$$

From the units alone, you can determine that the second one is the correct way to calculate speed since miles per hour is a unit of speed and hours per mile isn't. But let's take a look at this more closely and make two observations.

First, it should be noted that the first equation does give a correct number – you are indeed traveling such that it will take 0.2 hours (12 minutes) to go one mile. This is not speed because speed is expressed in distance per time, such as miles per hour. But it is true that you are going at a rate of 12 minutes per mile. This is also useful! I hasten to point out that the professor will mark it wrong, though, since this isn't what she was looking for.

Second, if the units are correct, there is a good chance that the equation is correct. It is no guarantee – perhaps you thought the equation was

$$speed = \frac{2.54 * distance}{time} ? = 2.54 * \frac{15\ miles}{3\ hours} = 12.7\ \frac{miles}{hour}$$

If the units are wrong, you know the equation is wrong. If the units are correct, you know you have a good candidate for a correct equation. If you do not know the equation, then, sometimes you can find good candidates just from the units and then use other methods to determine which candidate is correct. If you have a multiple choice test, for example...

Imagine the delight Einstein must have experienced when he first theorized that mass could be turned into energy, theorized that the equation would be $E = mc^2$ and the units worked out. Oh what a feeling that must have been.

Or perhaps it was the other way around. Perhaps he was playing around with units and noticed that the units worked out such that one should be able to mathematically go between mass and energy.

If any reader knows what role units played in Einstein's discovery of $E = mc^2$ or similar discovery, please post the story in the group discussion as described in the *Introduction*.

What do you think?

Why do you have to keep your eyes closed? Because otherwise, you would be an observer and the pet would become either a dog or a cat. Or perhaps the universe would split into a universe with you and a cat and another universe with you and a dog. But either way, for the pet to remain undetermined, you have to refrain from observing or interacting with the pet in any way.

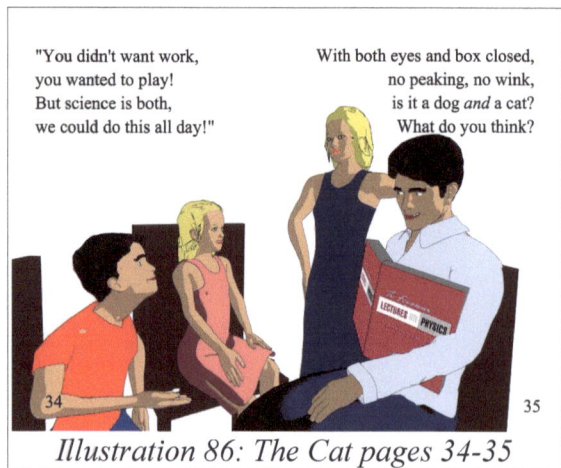

"You didn't want work, you wanted to play! But science is both, we could do this all day!"

With both eyes and box closed, no peaking, no wink, is it a dog *and* a cat? What do you think?

Illustration 86: The Cat pages 34-35

The series logo is in the Tim's hand.

What's next?

And so ends this first book in the *Mom, I wanna be a Scientist* series. We hope that you both enjoyed the story and learned to accept wave-particle duality and other concepts presented in this story.

In the next story we'll explore the concept of relativity. One interesting thing that relativity teaches is us that time slows down for objects that are

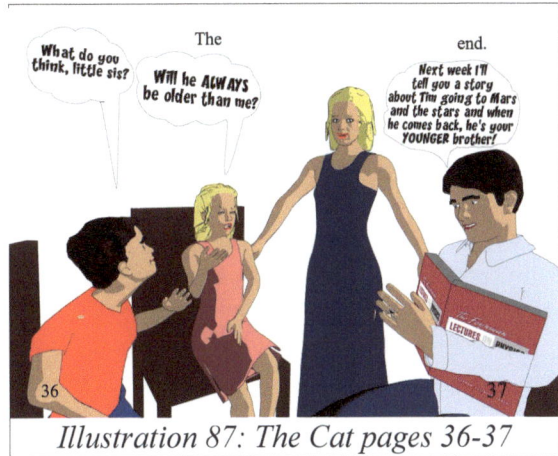

Illustration 87: The Cat pages 36-37

moving. If you were not exposed to this concept at an early age this may seem strange to you, but it has been tested many times by many different people since Albert Einstein discovered it in 1905.

If Tim travels fast enough and then turns around and comes back, time will not pass for him as quickly as it passes for his stationary sister. If he travels fast enough and long enough, turns around, and comes back to meet his younger sister, he will find that she is older than him.

This theory has been tested over and over and every experiment agrees with the theory of relativity. Take that, brother, and may it serve you well.

The series logo is on Dad's left hand ring finger.

The Lost Pages

The remainder of this book includes lost pages and their accompanying explanations.

Some are lost pages because they were deemed non kid friendly, for example, the Schrödinger and Einstein's original Schrödinger's Cat story kills the cat (well it *might* kill the cat).

Some are lost pages because they were deemed too complicated for the children's story, for example Wheeler's delayed choice experiment. Others were taken out for various reasons and I wanted to include them here because either they gave me an opportunity to give an important explanation that wasn't in the book pages or simply because I thought they were fun.

While these pages are not important to the children's story, they are important for a thorough introduction to modern physics. If you enjoyed the previous part of this book, you will probably enjoy this section at least as much.

Could be aether

Near the end of the 19th century, it looked like physicists were getting close to knowing all the secrets of the universe. One little detail they had left was to measure the speed of light. They "knew" light was a wave (ha!), and waves travel through some sort of medium – ocean waves travel through water, for example, and sound waves travel through air. Although no one could see or otherwise detect the medium through which light traveled, they assumed there must be a medium and called it aether (pronounced "EETHER"). The concept of aether was widely accepted at that time. In the 1880's Albert A. Michelson and Edward W. Morley made very careful measurements to measure the speed of light through the aether and to determine the speed of the earth relative to the aether. To their surprise, they were not only unable to find the aether, but they also found that the speed of light was a constant independent of the earth's speed and in the beginning of the 20th century the idea of an aether was abandoned.

Illustration 88: Original picture for Sam's book The Cat in the Box

This was the original picture for the cover of the book. The aether joke was deemed too esoteric and distracting for the audience. We tried several iterations, including using the alternate "ether" spelling, but none flew with proofreaders.

There has been some recent indirect evidence that there is a medium similar in concept to the late 18th century aether. As of the beginning of the 21st century, then, I can hear the aether quietly screaming, "*I'm back....*"

The birth of relativity

Suppose Tim is speeding down the street. Standing still on the street, Dad points a radar gun at Tim and measures his speed to be 3 meters per second (mps). We say "Tim's speed relative to Dad is 3 msp."

Sally is driving behind Tim. Mom is standing still on the street with Dad, and measures "Sally's speed relative to Mom" as 1 mps.

Quiz: What speed will Sally measure for "Tim's speed relative to Sally"?

Classical physics tells us that Tim's speed relative to Sally will be 3-1=2 mps. If you set up this experiment, this is what you will measure.

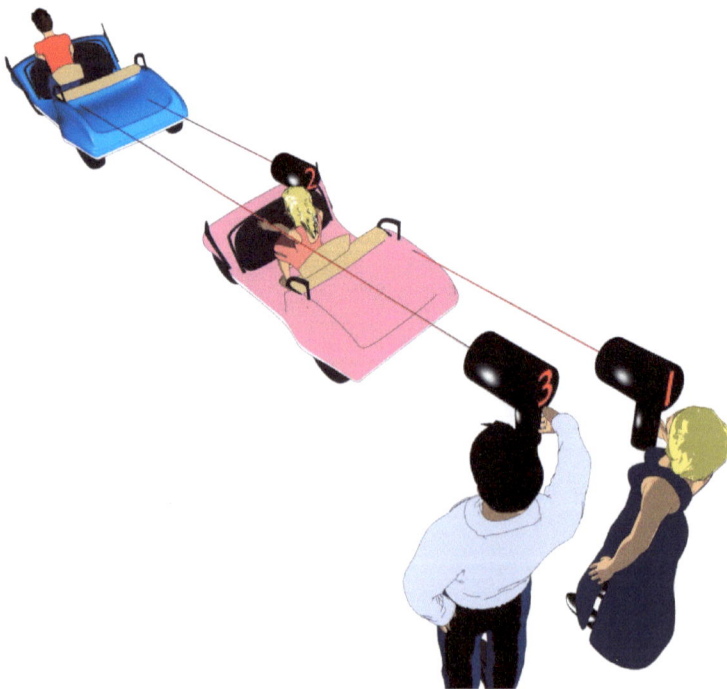

Illustration 89: Relative Speeds

But if Tim is traveling fast – very fast – this is not what happens. Let's see what happens when Tim and Sally travel a hundred millions times as fast, putting Tim at the speed of light. What Michelson and Morley found was in their experiment was that if we replace Tim with light, Mom and Sally will measure the speed of light to be the same value relative to themselves, even though relative to Mom Sally is traveling in the same direction as the light.

So let Tim get out of the car and shine a light (shown in violet, but the color doesn't matter). So Dad measures the speed of light relative to Dad to be 300 million MPS. Mom measures Sally's speed relative to Mom as 100 million MPS. Most of us expect Sally to measure Tim's speed relative to Sally as 200 million MPS. But that's not what happens. Sally measures the speed of that same light relative to Sally as 300 million MPS.

This was the surprising result from the famous Michelson-Morley experiments in 1887.

I want to note that this is very much simplified. A radar gun will not measure the speed of light, and much more sophisticated arrangements are used in the actual experiments. For an appreciation, read up on the Michelson-Morley experiments.

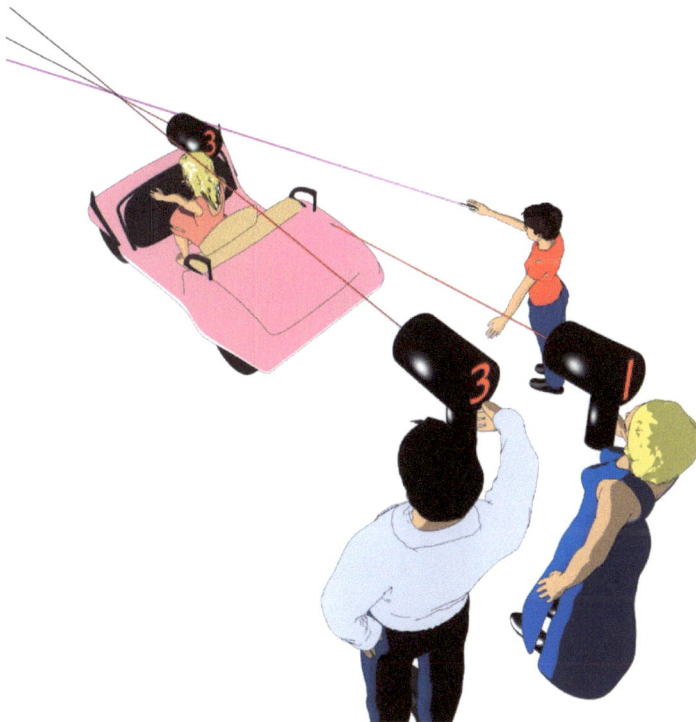

Illustration 90: A really really simplified version of the Michelson-Morley experiment

This seemed bizarre, nonsense, and impossible to most the physicists of the day (who had not read the *Mom, I want to be a Scientist* series as children :) because the results of the experiment disagreed with any sensible math they applied to it. By "most," I mean "all but one." The lone physicist was Albert Einstein.

While most of the physicists were immobilized by the Michelson-Morley results because they could not apply any sensible math to the problem, Einstein took another approach. He took the results of the experiment literally and followed the math wherever it took him, even though it did not appear to be sensible. He kept following it and developed an entire theory of what happens if one accepts that the speed one measures for light is independent of one's motion. He called this the theory of relativity.

He then made predictions based on this theory. The predictions he made, for example that time slows down and that objects get shorter when they travel fast, were so non-sensible that almost no one took the theory seriously.

But over the next few decades people started doing experiments to test Einstein's predictions. The results for every experiment turned out to be exactly as Einstein predicted and slowly – over many years – people started accepting relativity.

Armed with a quantum understanding of the photon, electron, and proton along with the theory of relativity, their confidence grew during the first quarter of the 20th century. It looked like physicists had figured everything out there was to know. They had a few things to tidy up and measurements to make to get more precise values for constants, but these were just details. Many became confident that discoveries were just about complete. In fact, in 1928 physicist and Nobel Prize winner Max Born told a group of visitors to Gottingen University, "Physics, as we know it, will be over in six months." His confidence was based on the recent discovery by Dirac of the equation that governed the electron. It was thought that a similar equation would govern the proton, which was the only other particle known at the time, and that would be the end of theoretical physics.

Not long after that, however, the neutron was discovered along with a plethora of subatomic particles that proved Max Born wrong on this particular point. *Very wrong.*

After reading the last section, one of my friends asked, "Really Jerry? *Plethora* of subatomic particles? Are there really that many?" I referred him to a famous quote by Enrico Fermi mentioned in the earlier section *Atoms*. When one of his assistants corrected him on the name of the one of particles, his reply was "Young man if I could remember the names of these particles, I would have been a botanist."

Experts don't know everything

Max Born is not the only expert to miss the boat on predictions in his field. Dominick O'Brien[9] notes his top five blundering quotes from experts who it seems in hindsight should have known better:

1. "Who the hell wants to hear actors talk?"
   - H. M. Warner, President Warner Brothers at the introduction of movies with sound, 1927

2. "I think there is a world market for maybe 5 computers"
   - Thomas Watson, Chairman of IBM 1943

3. "This telephone has too many shortcomings to be seriously considered as a means of communication. The device is inherently of no value value to us."
   - Western Union internal memo, 1876

4. "Heavier than air flying machines are impossible"
   - Lord Kelvin, president of the Royal Society, 1895

5. "Drill for oil? You mean drill into the ground to try and find oil? You're crazy!"
   - Drillers employed by Edwin L. Drake, founder of the modern oil industry, 1859

Experts don't always know everything.

Here is something to think about: would applying the scientific method to O'Brien's quoted problems have led to better predictions? Application of the scientific method would have caused market surveys to be performed on the first three of O'Brien's quotes because these are marketing questions. Though I have not tested this theory in any way, I believe application of the the scientific method would have caught these three mistakes. The last two O'Brien's quotes are scientific questions and application of the scientific method would have caught the mistakes. Consider the last quote. The scientific method <u>was</u> employed. Drake *did* test out the theory that if you drilled into the ground you could find oil.

---

9   In his memory course, *Quantum Memory*. Note that his technique has nothing quantum about it. The term quantum has been repeatedly abused. One of the top battery companies just came out with a line named quantum, which has no more to do with quantum than any other battery. Nothing quantized here!

John Wheeler's delayed choice experiments

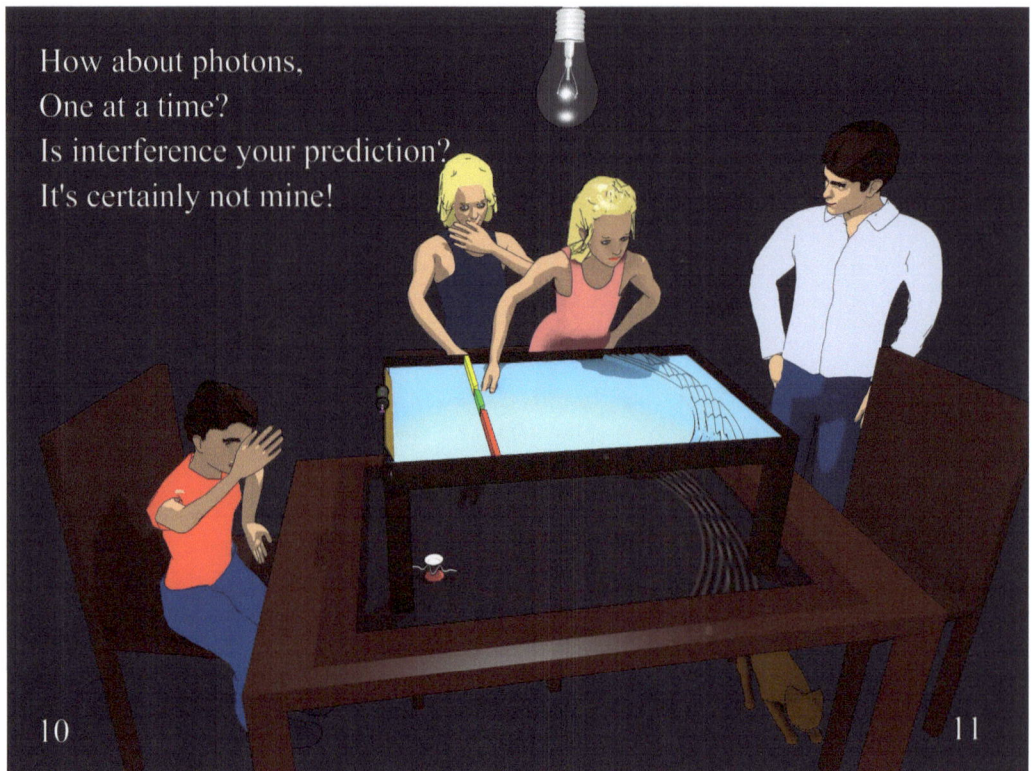

Illustration 91: Sending one photon through at a time

Illustration 91 was an introduction to the brilliantly clever experiments of John Wheeler. The concept was deemed too advanced to include in *The Cat in the Box*. It is an important concept and provides insight to key concepts of modern physics, however, and I include it here.

When an object such as light travels through slits, its propagation can be described as a wave. When it interacts with a detector or anything else, it can be described as discrete particles. We now discuss an interesting question: when and where does a particle model serve us better and when and where does a wave model serve better? Or, using the experiments we have been discussing, when and where will observation show an interference pattern? Beginning in 1978 John Wheeler proposed a clever series of experiments to answer this question.

The series logo is on the left side of the table under the ripple tank.

How we use science – Determination versus demonstration

Before we dive deep into John Wheeler's experiments, it's worth a detour into models and experiment controls to make sure we are using the power of science to honestly find what we're looking for.

Using Statistics to estimate customer preference

Suppose that I am an statistician employed by the engineering department of a toothpaste manufacturer. Our scientists have come up with five new formulations of toothpastes and I have been tasked with using surveys to find out which ones the majority of our customers think tastes best.

Here is what I do. I decide how many people I'm going to survey. It's too expensive to ask all our customers, so I need to choose a small sample of customers who represent the entire customer base. I'll use this to estimate what the entire customer base would prefer. I need to make sure I survey enough customers – if my sample size is too small, that's not going to give me very much confidence that my answer is going to represent all my customers very well. For example, if I just sample one person and that persons happens to hate mint, I'm not going to get a very high quality estimate.

So let's say I chose 100 people. Next, I need to make sure it's a fair sample of the population. Suppose I choose 100 children and they choose the sweetest tasting one but children only represent a small percentage of our customer base. This is not going to give me a good estimate either.

I also have to be careful not to influence the people taking the survey. I do not want to tell them, for example, that sample one has natural cinnamon flavor and sample two has artificial cinnamon flavor because I'm trying to determine which one they think tastes better, not their opinion of whether we should be using natural or artificial flavoring. So if the packages are labeled, I would blindfold them for the survey. Another way to do this is to put them all in identical unlabeled packages. I would then label the containers somehow that doesn't give away any information, such as "Toothpaste A", "Toothepaste B", etc. Either way, as long as they don't get any information other than the taste, this is called a blind test.

But I have to take this one step further. Suppose I myself have strong feelings that natural cinnamon is superior to artificial flavoring. I might without knowing it influence their opinion. I might smile when giving them the natural version and grimace while giving them the artificial version without realizing I'm doing it. A solution is to have someone else put them into the containers without telling me which one is which. That way, I cannot give the survey takers any information. Since the survey taker does not know which is which and I do not know which is which, this is called a double blind test.

This is going to a lot of trouble, but if you are trying to get an honest answer to the question of which tastes best to our customers, you have to do it this way.

I'll do several different surveys. If most of them agree with each other, then these results are likely to be a good estimate of what my entire customer base prefers.

I need to make sure I didn't influence the results in any way. For example, if it turns out that I had labeled them #1, #2, #3, etc.) and no matter which one was put into which container customers always preferred #1, I have to find an alternate labeling. We would find this out using experiment controls which I'll talk about shortly.

I must include every survey I did in my results. If I'm honestly trying to determine what our customers like best, I cannot do ten surveys and report only the ones that came out with the result that agrees with my personal preference. Even if the test is messed up because of labeling, for example, I still include that test when I report the results, but I explain in my report that this test was messed up and why.

How to lie with Statistics

Now suppose that my brother is a statistician employed by the marketing department of the toothpaste manufacturer. Our scientists have come up with new formulations of toothpaste and he has been tasked with using surveys to demonstrate that the majority of consumers think our new formulation tastes better than our competitor's product. His boss in the marketing department tells him it would be nice if three out of four customers preferred our product.

Here is what he does. He chooses a sample size of four. He randomly chooses four people from different states (Say Kentucky, Florida, California, and Wisconsin) who do not work for the toothpaste company and gives them the same carefully planned double blind test that I did. Suppose half of them prefer our brand. That's survey number 1. Then he finds four more people and repeats the survey. In Survey#2, none of them preferred our brand. So he finds four more people. In survey#3, three out of the four preferred out brand. He then carefully documents the third survey, noting that it is a random selection of people who do not work for the company. He notes that it was a double blind study and anything

96

else that would make this survey appear to be a good scientific survey.

The marketing department is thrilled with his work. They create a national commercial boasting that "In a national survey, three out of four people preferred our brand." He publishes a paper describing how the survey was done. If their sales go up, he gets a raise.

I am not claiming that this is unethical. There was no actual lie told – the survey was done. As are all surveys by our company, survey #3 was done fairly, and the results of that survey were reported accurately. There is no law that says you have to have a sample size large enough to represent the population. There is no law that says you have to have report every survey you do.

Both sets of surveys accomplished their goals. But there goals were *very* different. The first set had the goal of <u>determining</u> the preferred product. The second set had the goal of <u>demonstrating</u> that a particular product was preferred. You have to clearly understand what it is you are trying to accomplish and then honestly set out to do it. If you are the recipient of test data, you had better understand who paid for the tests and what their goals were.

Experiment Controls

Sally learned in school that not all animals have their ears located on their heads. Suppose Sally wants to determine what body part of a frog contains its ears. She comes up with the theory that a frog's ears are located on their hind legs and wants to test that theory.

So what she does is teaches the frog to jump when she says the word "jump!," giving him a nice juicy fly every time he does it correctly. After a few days of this, the frog has learned the behavior and jumps every time he hears the word "jump!"

Next, she cuts his hind legs off and says "jump!" The frog doesn't jump and she concludes that the frog could no longer hear the command. Clearly, she concludes, frogs' ears are located on their legs.

What did Sally do wrong? She changed something in the experiment without really understanding that it would take away the frog's ability to jump.

When doing experiments, it's important to understand all the effects of each thing you're doing. One common approach is to repeat the experiment a number of times, changing only one thing at a time[10]. The first form of the experiment should not be testing the theory, it should be confirming the setup to make sure that you understand everything about it and that no mistakes have been made in the experimental setup. This test of the apparatus and understanding of the experiment is called a control.

Here's one way to introduce a control into the experiment. Because a frog's eyes are easily visible, you know where they are – they are on his head. So you train the frog to jump not only when he hears the word "jump!" bit also when he sees a flash of green light.

Next, you confirm that he consistently jumps for either the word or the flash. Then you cover his eyes and confirm that he jumps when you say "jump!" and not when you flash the light. Then you uncover them and repeat the experiment. Then you cut his legs off and say "Jump!" He doesn't jump. But then you repeat the control – you flash the green light. If he jumps with the flash and not with the word, then it is a reasonable conclusion that his ears are in his legs. But if he doesn't jump with the flash, then you have not confirmed your theory – something other than not hearing must be preventing him from jumping.

Richard Feynman did a great job summing up the importance of experimental controls in a commencement speech he gave at Caltech in 1984 where he describes a study done on rats in a maze in 1934. This speech, called *Cargo Cult Science* can be found on the Internet and is well worth the read. If you search videos, you can find people reading the speech if you'd rather listen to it. It's also included in his popular book *Surely, you're joking Mr. Feynman*. This is a must read.

Well, that's not very kid friendly. Nor Frog friendly!

I could have come up with a less frog-traumatic example. But this is a classic story familiar to many people and it does illustrate the point well. This is the classic way of telling the story. If you're looking for a more kid-friendly version, you might consider telling the story by covering the frog's legs with tape rather than cutting them off.

_____

10 Another approach is to change many variables and measure average differences. This is common when there are many complicated variables and it is impractical do to enough tests to vary each variable independently sufficiently to cover all possibilities.

Software Controls

I use this very same technique when I write software. Suppose that I need to compute the square root of 625. Realizing that I'll probably need to find the square root of lots of numbers, I'll go ahead and write some software to find the square root of any number.

My theory is that SQR will take the square root of a number. My program is very simple:

N=625
Answer=SQR(N).
Print Answer

Since I don't know the square root of 625, I don't know if this program gives the correct result. So before relying on it, I run it with some numbers I already know the squre root of. Let's try 9:

N=9
Answer=SQR(N).
Print Answer

When I run the program, I get 81! Well, that's not right. Perhaps SQR means square and not square root. So I run it with 4 and get 16. And I run it with 16 and get 256. This software control, where I know what the answer should be, caught an error. Perhaps I need to use SQRT instead of SQR:

N=9
Answer=SQRT(N).
Print Answer

This program returns 3. N=16 returns 4. N=100 returns 10. Now I have confidence in my program and I can run it with N=625. That gives me an answer of 25 and I am reasonably confident that the square root of 625 is 25.

This is a great technique to use when you're taking a test in school. If you have a problem and you *think* you know the equation or formula but are not sure, do a similar one that you know the answer to and see if it comes out correctly. If it does not, you may want to take a closer look at your formula.

John Wheeler's delayed choice experimental setup

The test that John Wheeler devised was to repeatedly shoot a photon through a multipath arrangement such as the double slit experiment and watch for an interference pattern. He would repeat the experiment over and over, blocking one of the slits at various times in the experiment and see if it resulted in an interference pattern. Recall from the wave behavior section the presence of an interference pattern indicates wave behavior. By determining what parts of the path would disrupt the interference pattern, he could determine when and where the photon's path is described by wave equations and when and where its path is described as a particle. It was a genius idea.

The experiment goes like this: repeatedly shoot some light in an arrangement where the light could go through two or more possible paths such as slits. Wait until the light has definitely had time to go through the decision point between the paths but not had time to make it to the detector. Block one path at this time. When the photon gets to the detector see if there is an interference pattern.

There are many variations on the delayed choice experiment. They are all difficult to perform because light travels so fast and closing the slit while the light is en route is extremely challenging even with the latest technology.

The delayed choice experiments that are practical enough to be done in a laboratory are difficult to understand because sophisticated equipment is required to turn the switch on and off sufficiently quickly.

Unfortunately, no one has developed a delayed choice experiment that can be done in your kitchen. Therefore, we will discuss a theoretically possible version which is easy to understand but is not practical to do in your kitchen.

After Wheeler proposed the experiment, but before it had been performed, everyone seemed to think they could predict the results of the experiments. But people's predictions did not all agree. The response to Wheeler's predictions, according to Ross Rhodes[11], "ranged from *Nonsense, Rubbish,* and *Completely* absurd to *Yup, of course."* So this was going to be an interesting experiment.

Although the concept is elegantly simple, performing the actual experiment was anything but simple. The problem is that you need to place obstacles into and out of the paths while a photon is traveling through the apparatus and the photon is

---

11 Ross Rhodes is a science writer and lecturer for both professional and lay audiences, specializing in the philosophical implications of quantum mechanics.

traveling so very fast, it's nearly impossible to place obstacles in and take them out at just the right time. So Wheeler and others devised clever but complicated arrangements to accomplish the tests. What I'll discuss is a simplified version useful for explaining the concepts, but the actual arrangements done in the lab are more complicated and sophisticated to meet the challenge of such short time frames. We'll imagine doing each experiment with both water waves and photons.

For Experiment WDC1 (this stands for Wheeler's Delayed Choice 1), we'll repeat the usual double slit experiment to check the apparatus and serve as a control. We turn on the wave source and watch for an interference pattern.

As expected, an interference pattern shows up at the right side of the apparatus for both water and light.

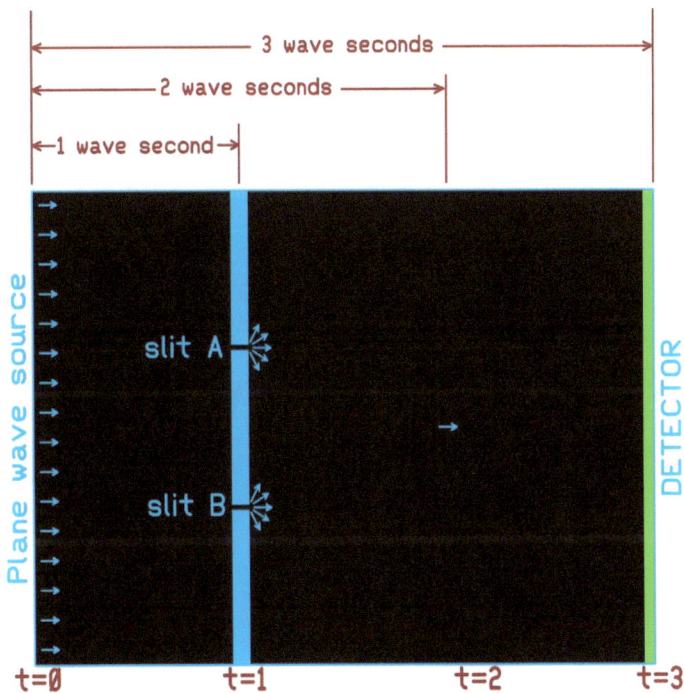

Illustration 92: Repeating the double slit experiment as a control

Next, for WDC2, we'll repeat the same experiment shooting one photon or a water wavelet through at at time. This is the experiment done on pages 12-13 in *The Cat in the Box*. After repeating this for many photons, one at at time, and adding the results, we see an interference pattern. This confirms that one photon "interferes with itself" in this apparatus.

For WDC3, we'll block one of the paths by closing slit B one second before we start the experiment. We then send photons through and find that there is no interference pattern. This is expected because there is only one possible path available for travel. This is the experiment shown in Illustration 91 earlier in this section where Sally is blocking the path of one slit.

In these experiments so far we find that interference occurs when and only when both paths are available whether we send many photons through at a time or we send them one at a time. The photon particle seems to "know" when multiple paths are available. But remember, the photon is not a sentient being. It doesn't *know* anything.

Wheeler's question was: when/where (remember, time-space and all) does the photon determine if both slits are open? Classical reasoning says it's when it goes through the slits. *Of course*, right?

The next experiment, WDC4, is to send one photon at a time and close slit A while the photon is traveling through the apparatus.

To make sure we're on the same page, let's add a time description to the experiment. The apparatus is three feet long and it takes light about one nanosecond (ns) to travel one foot. So it takes light 3 ns to travel from the left side of the apparatus to the detector at the right side. So, at t=0 ns we send out a photon. The photon arrives at the slits at t=1 ns. At t=2ns the photon is halfway

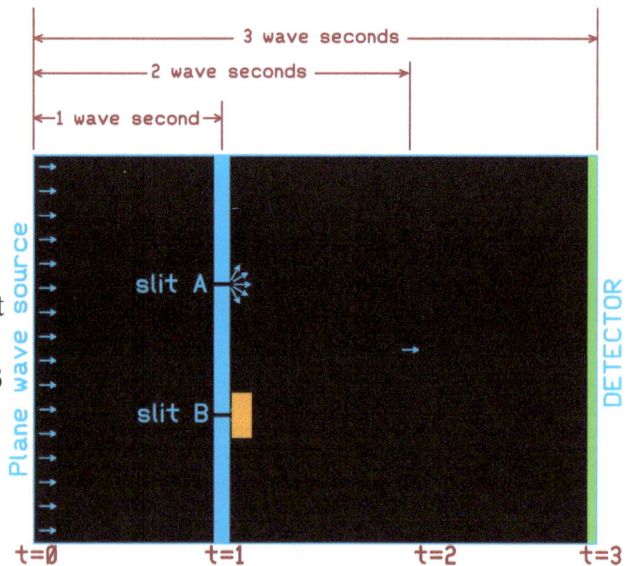

Illustration 93: Blocking one of the slits

between the slits and the detector. The photon reaches the detector at t=3ns.

What Wheeler proposed was sending the photon into the apparatus at t=0, waiting for the photon to go through the slits, let it travel toward the detector, close slit B at t=2ns, and look at the detector at or after t=3.

As mentioned earlier, everyone seemed to think they could predict the results of this experiment. But people's predictions did not all agree. So this was going to be an interesting experiment. Before reading on, take a break from this book and discuss it with friends. Perform the experiment with water on a wave table if you have one. Discuss it with as many people as you can over the next few days before reading on. If you discuss it with at least five people, you're very likely to get differing predictions.

Don't just read on, go discuss! Make predictions and record them here:

_____

_____

_____

_____

_____

_____

_____

_____

Illustration 94: Blocking the slit after the photon goes through it

Illustration 94 shows the concept of Wheeler's delayed choice. Sally is blocking one of the slits after the photon has passed through the slits.

It's time for the crux experiment WDC4 where we add the delayed choice twist. We turn on the wave generator for a short time and send down a wavelet of water or a single photon. The photon travels from left to right. AFTER the photon or wavelet goes through the slits but BEFORE the it hits the right side, we block the path of one slit and watch the right wall. This is what is going in Illustration 94: Tim sent out a short burst of waves and the burst traveled through the slits. Sally then blocks one of the slits. You should try this! When you do this, the wave has already traveled through the slits so *of course* you get an interference pattern.

But with light, this doesn't happen! If you close the slit after the light goes through the slits but before it hits the detector, the interference pattern fails to show up! Even if you put the obstacle in at 2.9999 ns – long after it has gone through the slits, there is no interference pattern. The light seems to "know" that the slit was closed after it went through it. A popular explanation is that the photon travels back in time to tell itself whether to go through one or both slits. Another

104

explanation is that the particle has an extensive wave field that is spread out in space. Another explanation is that Martians come down and pull the photon back. There are lots of explanations proposed. Many different models can be useful for different purposes. The scientific method can be used to choose between them. The result of this delayed choice experiment is one of the bizarre effects of quantum physics that stymies many people. But those people haven't read this book :)

This is counter-intuitive to everything we have observed in our life because the big things we observe on a daily basis don't appear to demonstrate wave characteristics. But little things *do* follow wave equations. This experiment has been run with a variety of objects in addition to photons. This works with electrons. It works with helium atoms. It works with buckyballs. They all fail to show interference patterns if the path is disrupted before the photon hits the detector. There is no interference pattern with the disruption, even if the disruption is delayed until after the object has passed the slits[12].

What I'd like you to do now is go back and re-read *The importance of diagrams,* the very first explanation in this book, the explanation pages for pages 2-3 that describe how to use algebra to find the solution to the question of where to draw the explosion on the video game. If you read it and the next few pages carefully and take it seriously, Wheeler's delayed choice results will not seem radical or surprising.

About that video game

How did you propose to solve the laser explosion problem? One way that occurs to many people is to start tracing the path of the laser beam when the user pulls the trigger and watch for it to hit he wall. You start out at, say, T=0.01s and compute whether it has hit the wall yet. It hasn't. So then you try T=0.02s. It still hasn't hit it, so you try T=0.03. When you get to T=0.424s, you see that the laser beam has hit the wall and you compute X and Y. This is a time-based solution and is natural to many people because they are solving it in the order things happen because they find it intuitive based on their experiences.

Notice that in the given solution, we did it very differently: we chose to find time at the end – we first find X and Y, then find T. We solved the system simultaneously without treating T as ordered or special in anyway – it's just another dimension along with X and Y.

---

12 This isn't necessarily true for <u>all</u> objects with <u>all</u> amounts of delay, as we will see later in a blue page but this is a detail to be ignored for now.

A different view of time

In order to understand the results of Wheeler's delayed choice experiments, let's give up the notion that time is ordered and think of it as just another dimension. In the video game laser game example that's just what we did – we solved for all possible solutions in a kind of batch process without ordering x, y, or T. Like Einstein, we just followed the math and found a solution that satisfied all the conditions without regard to what makes sense from our regular experience of things happening in order. Developing the ability to think this way is absolutely essential for both of the major components of modern physics, quantum physics and relativity. This is hard for an 18 year old – it takes maybe three or four years to get a rough handle on it, and many years more to become truly comfortable. But if done from a very early age, that investment will give a native, intuitive understanding for life.

Quick summary of Wheeler's delayed choice experiments

We turn a light on for a short time creating a pulse of light. The light travels through a double slit, going through one or both slits and then hits the screen. Thus far, we are simply repeating the double slit experiment with a pulse of light and we get an interference pattern. We repeat this with shorter and shorter pulses of light until we're sending one photon through at a time. We add the results of many runs on the single photon and get an interference pattern.

We now add the delayed choice twist. Suppose that the screen is far away from the slits, so there is some appreciable travel time between when the light goes through the slits and when it hits the detector. We shoot the light through the slit(s) and wait until the light is almost at the detector. We then close one of the slits before the light hits the detector, but long after it has gone through the slits. The seemingly bizarre result is that there is no interference pattern! A popular explanation is that the light seems to "know" that the slit was closed after it went through it. It seems to travel back in time to know this. Let's take a closer look before we make such broad claims.

Let's take a step back and review some of the concepts discussed earlier.

Everything exhibits some wave properties and some particle properties. The smaller the object, the easier it is to observe the wave aspects. Additionally, the slower something is moving, the easier it is to observe its wave aspects. A wave equation can be associated with any object (a photon, a proton, a cat...). What that means is that we can find the equation of a wave to describe the object such that the wave going through any path predicts the outcome of an experiment of that object going through the path. Let's take a look at what some of these wave functions look like.

Wave functions and probability functions

Everything has a wave function for a given situation. Light going through a slit has a wave function. An electron traveling from a light bulb has a wave function. My dog chasing a squirrel though my yard has a wave function.

Let's take a look at a wave function. Illustration 95 is a sample wave function. The derivation of this function – how we arrived at this particular function is covered in an advanced physics course.

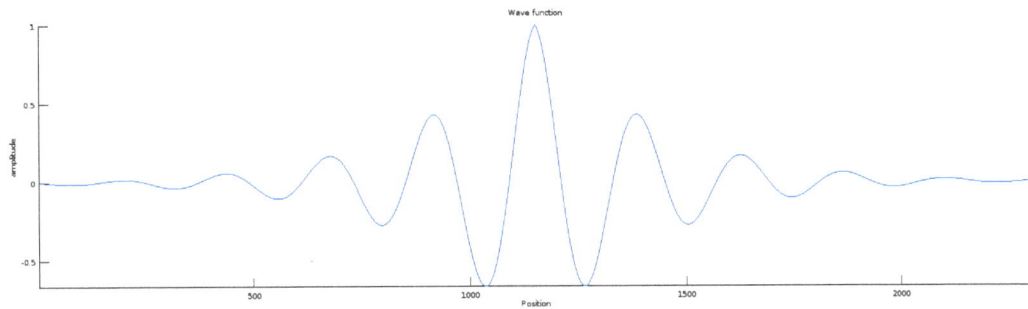

Illustration 95: A sample wave function

A common use of the wave function is to calculate a probability function associated with the particle. This is done by squaring the wave function and for this wave function the probability function in Illustration 95 looks like Illustration 96. This probability function tells us the probability of finding the particle if you look in a particular place. This particle does have a location which is more probable than any other location and we'll call this the most probable location. But it might be found in other places. For example, although it is unlikely, it is possible that it would be found at position 1500. There are places, however, that it will never be found. For example, it will never be found at location x=970 or x=1090. These correspond to wave cancellation in the same way that there were places in the ripple tank where the water was calm and places in the yard where no sound could be heard because the waves canceled at that location.

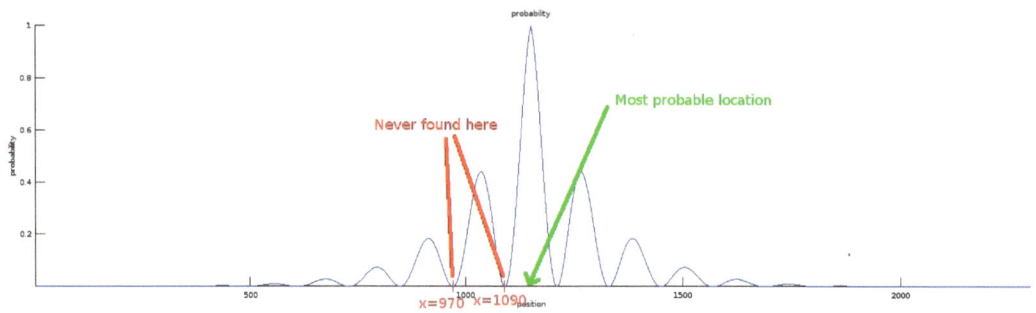

Illustration 96: The probability function found by squaring our sample wave function

Imagination and visualization

Pages 28 and 29 of Sam's The Cat in the Box shows a stamp with Einstein's famous quote, "Imagination is more important than knowledge." Why did he say this? He said this because in physics there are a lot of abstract concepts and math that you must somehow learn, understand, and use to take full advantage of the information. This can be challenging and it takes a while to learn to fully utilize the many different ways that have been developed to present data.

Illustration 95 and Illustration 96 show one dimensional representations for a wave function and a probability function. So imagine that there's a particle somewhere along a ruler. Illustration 96 Shows the probability of finding the particle anywhere along the ruler. But what if the particle isn't restricted to a straight line. What if it could be anywhere on a plane, for example a dot anywhere on a piece of paper. Suppose I tell you that the wave and probability functions are the same as in Illustrations 95 and 96. To visualize this, imagine that we've spun the function around in a circle centered at the peak of the graph. That's still pretty hard to visualize, and a drawing would be very helpful. We're going to need a second dimension for position. One way to do this is with a three dimensional plot called a mesh as shown in Illustrations 97 and 98. Note that in addition to using height as the third dimension to show the value of the function, these particular meshes also use color. Not all meshes show color.

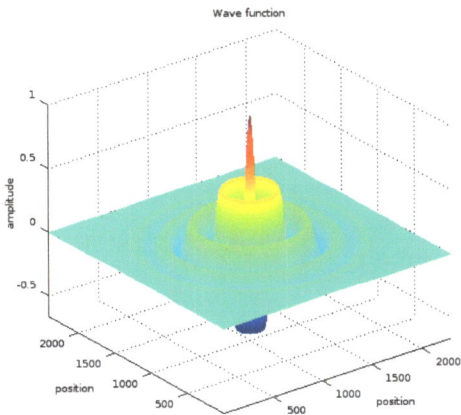

Illustration 97: Mesh of the sample wave function

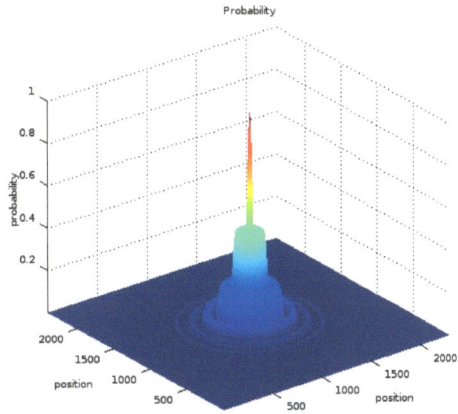

Illustration 98: Mesh of the sample probability function

This is not the only way to represent the data. There are many ways to present 3D data, each with at least one advantage and disadvantage. Suppose that there is a hidden feature behind one of the peaks. There might be a little bump that you cannot see in the back of the plot and you don't want to miss it! Another way to represent the data for visualization is shown in Illustrations 99 and 100 where we have used greyscale to show the value – the higher the value of the function, the brighter the function. Where it attains its highest value, we use white. Where it attains zero value, we use black.

Illustration 99: Greyscale representation of the sample wave function

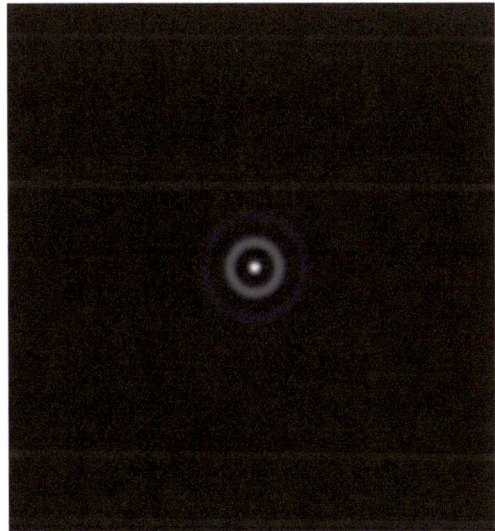

Illustration 100: Greyscale representation of the sample probability function

What type of representation you use depends on what you are trying to convey.

Most of the time you use the simplest one you can get away with, so if the concept can be conveyed using a one dimensional representation, that's what we use.

Some example probability functions

Here are some example wave functions. We begin with the wave function in Illustration 101 of something massive, say a dust particle. In this case, we know almost exactly where it is and it doesn't easily reveal its wave properties.

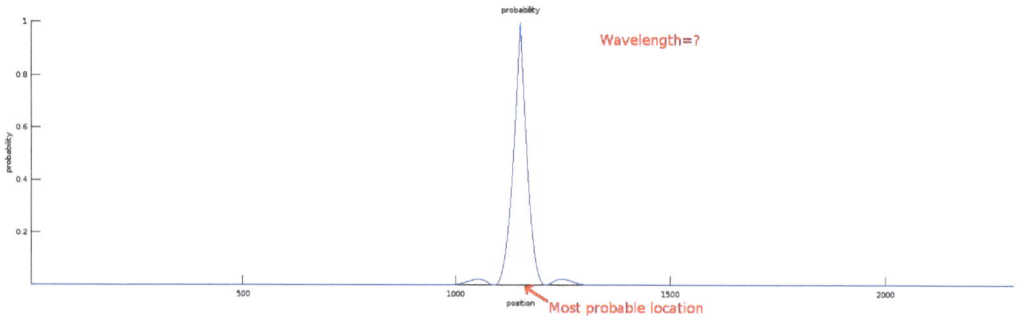

Illustration 101: Something with a lot of mass doesn't easily reveal its wave properties

Now we go to something with a lot less mass in Illustration 102, say a proton. Note that it's starting to show some wave properties – it doesn't have a definite position and we can start talking about a probable wavelength.

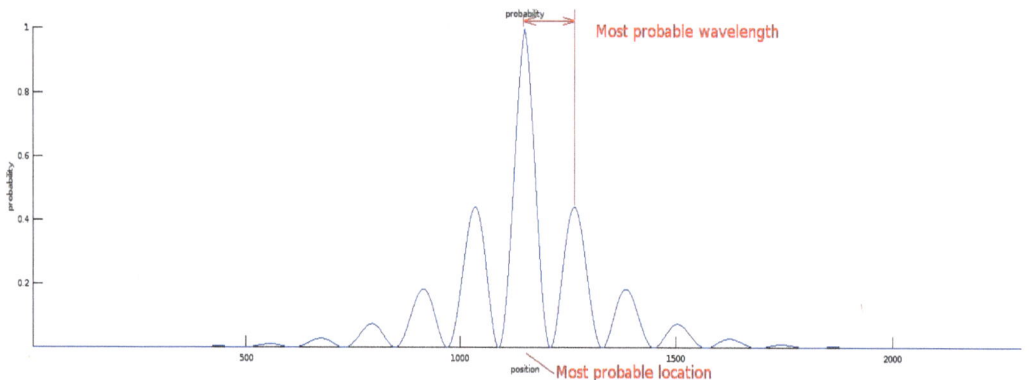

Illustration 102: Something with less mass starts to exhibit wave properties

In Illustration 103 we've gone to something quite small, say an electron. Note that its wave characteristics are really showing up. It has a very probably wavelength, but its likely position has spread out quite a bit.

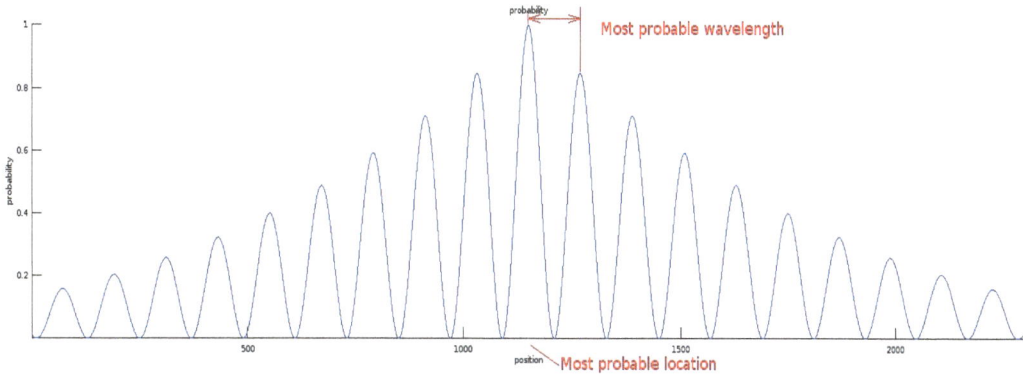

Illustration 103: Something with very little mass has a lot of wave property influence

In Illustration 104 we have gone to a completely massiveless particle, the photon. It has a definite wavelength but its position is very uncertain.

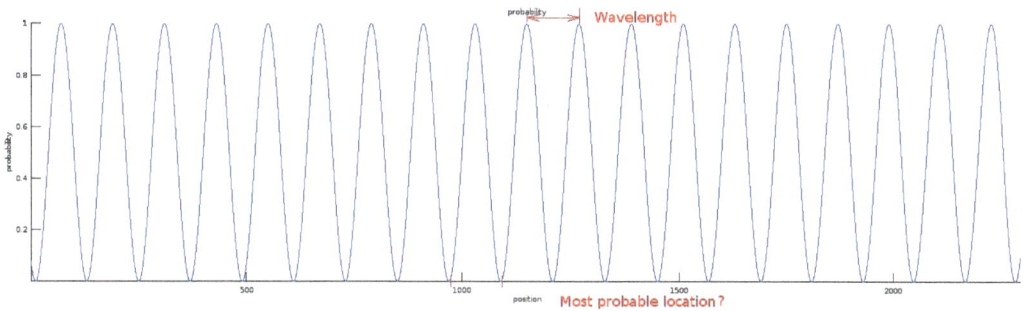

Illustration 104: Something with no mass that hasn't yet interacted with anything goes on forever

What on earth do these diagrams mean? Among other things, they show the probability that an object will be found if you look for it (i.e., put a detector) at some location. There will be somewhere where it's most likely to be found. Let's call that the "expected location."

So, if you were driving a car which has lots of mass and look to see where it is, you're almost definitely going to find it in the expected location. The car is large enough – it has enough atoms – that it follows Illustration 101 and the pulse is very narrow. This means that the car will be found where we look for it with practically no randomness in its location. This is why we can be sure we know exactly where the car is, despite the Heisenberg uncertainty principle: the car will be found at the expected location beyond any reasonable doubt.

But if we were looking for something much smaller, the probability of finding it somewhere else isn't so small. A buckyball's wave function, for example, follows Illustration 102. This is one of the main features of the small world that we don't experience in everyday life. Several books in the series will be dedicated to giving you experience with this. For example, we'll shrink down to the size of buckyballs and play hide and seek. That will be very interesting because the friends you're looking for won't be in a specific location. We'll hunt for tiny Easter eggs and the world's smallest afikomen. It's a little harder when they aren't really located in a specified place!

So the probability of finding something looks kind of like a fuzzy bulls-eye. Consider the following experiment: set a bunch of detectors, such as the sensor on a camera, and shoot a gazillion identical electrons in the center of the sensor. Even if they have the same path, they will not all be detected at the same place. The particles have the probability function show in Illustration 100.

The resulting pattern looks familiar. You might notice how these graphs are similar to the interference patterns we have seen. What can we learn from this?

It's not because the particles were in different places or on different paths or going different speeds. It's because *none* of them were in *any* particular place! They were all in the space and have a probability of being detected as shown in Illustration 100.

So this diagram can be interpreted in multiple ways. It can show the results of many particles shot at one time. It can show the probability of where a single particle is likely to be found. It can show where a particle is likely to interact.

Here's another way to interpret these probabilities. Suppose you interpret the pattern to be locations in time space at which the object has influence and can be influenced.

So let's redraw Wheeler's delayed choice experiment in Illustration 93, taking this influence interpretation literally. Since we're talking about photons, we would use the probability function from Illustration 104.

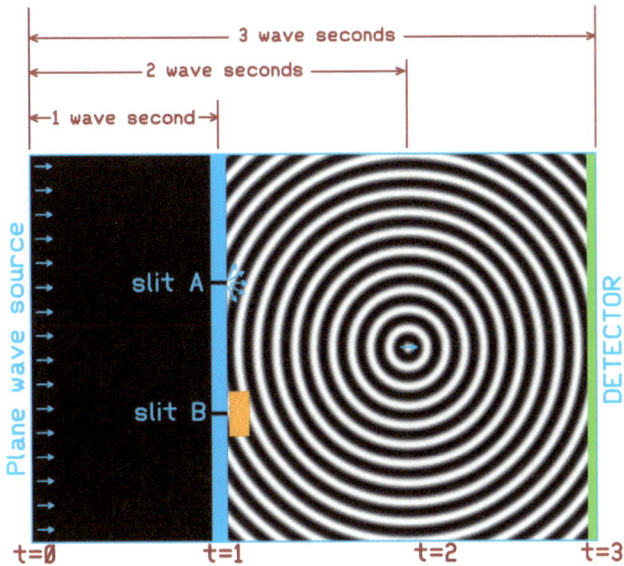

Illustration 105: Space of influence of a massless particle

So at T=2s, although the photon is most likely to be found halfway between the slits and the detector, you will see that the object still has some influence at slit B. Closing slit B at T=2s can influence its path because part of it is still there. Perhaps this makes the results of Wheeler's delayed choice understandable. Perhaps we have an explanation of what's going on. Do we understand it? If we wrote a theory based on this, made predictions based on this theory, and did the experiments, would the predictions be correct? Would the theory be true?

When is a theory proven true?

A scientific theory is *never* proven true. A theory makes predictions and then experiments are run to see if the results match the predictions. If the results do not match, the theory is dis-proven. If the results match, all that can be said is that the theory withstood that particular test. If many many tests have been done over long periods of time by many different people, the theory is widely accepted and has withstood the test of time. But it is not proven true – it can be overthrown at any time. The theory that heavy things fall faster than light things withstood the test of time for hundreds of years, as did the theory that light was a wave.

When is a theory useful?

In science, a theory is useful when it makes new predictions that can be tested and applied. So let's see what testable predictions we can make with this theory. Remember that we said the fall-off in probability has to do with the mass? Well, the mass of a photon is zero so its influence goes on forever. Therefore, let's make a testable prediction: it won't matter how far the photon is from the slit. If a slit is closed before the photon interacts with the detector, there will be no interference pattern.

Physicists can make make the apparatus longer, say as long as a football field, and it still works. How about miles long? This is been tested using miles of fiber optics in the path and it still works.

But I'm claiming it "goes on <u>forever</u>!" Is that testable? Astrophysicists did the experiment using a star as the light source and a black hole as a lens to provide two paths in lieu of slits. The choice of whether to look at the wave or the particle aspect was made more than a million years after the light passed through the decision point. When looking for wave characteristics, interference is present. When looking for particle characteristics, the photons arrive one at at time, in discrete chunks. The theory makes the correct prediction from three feet to a million light years.

While light is massless, an electron is not. Its wave function is like that of Illustration 103 rather than the photon wave function in Illustration 101. So we can make another prediction by asking ourselves what will happen when we do the Wheeler delayed choice experiments with electrons, which do not have an infinite tail like the photon. Think about for a while and discuss it with your friends.

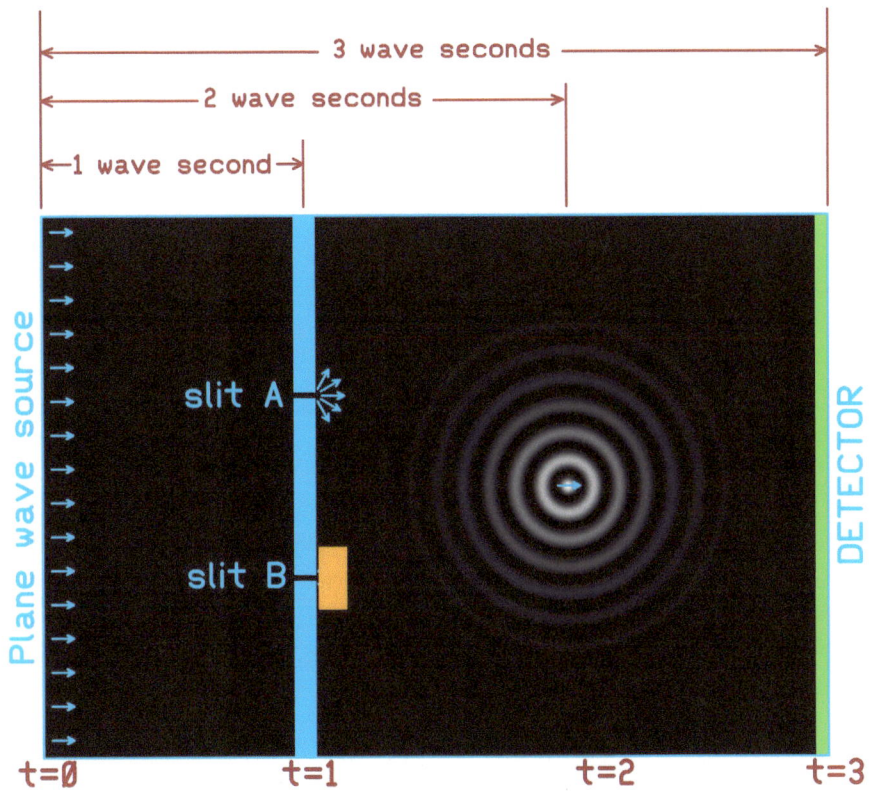

Illustration 106: An electron's probability function shown in Wheeler's delayed choice setup

Record your prediction here:

_____

_____

_____

_____

_____

_____

The prediction I make is:

A) when the detector is placed close to the slits and the electron does not have to travel too far between the slits and the detector, the tail of the electron will almost always be influenced by the slit and there will be no interference pattern. This is the case in Illustration 108.

B) when the electron has to travel very much farther than the effective length of the tail, the tail of the electron will almost never be influenced by the slit and there will be an interference pattern. This is the case in Illustration 108.

C) if you repeat the experiment at distances between A and C, the interference pattern will slowly appear as the tail has lower and lower probability at the slit. In Illustration 106, the tail of the electron is still in the slots, but at only about 10% probability. This theory, then, predicts that if you do the experiment in this set up, the interference will be present, but at only about 10% brightness. 90% of the electrons will show the particle clump pattern and 10% will show an interference pattern.

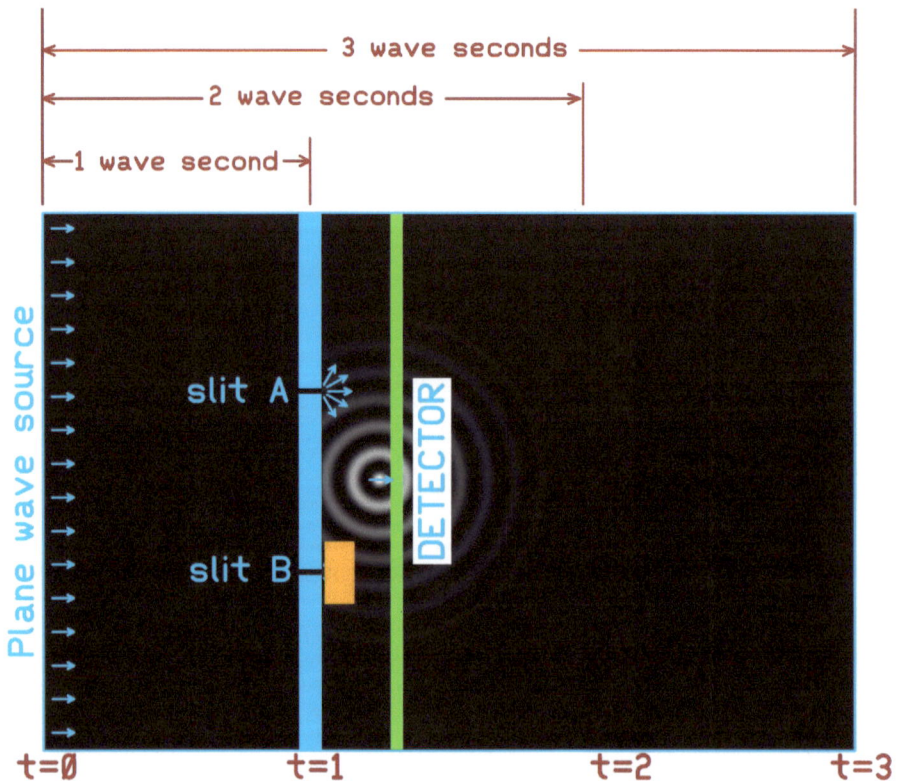

Illustration 107: Wheeler's delayed choice using electrons with the detector close to the slits.

3 wave seconds

2 wave seconds

1 wave second

Plane wave source

slit A

slit B

DETECTOR

t=0   t=1   t=2   t=3

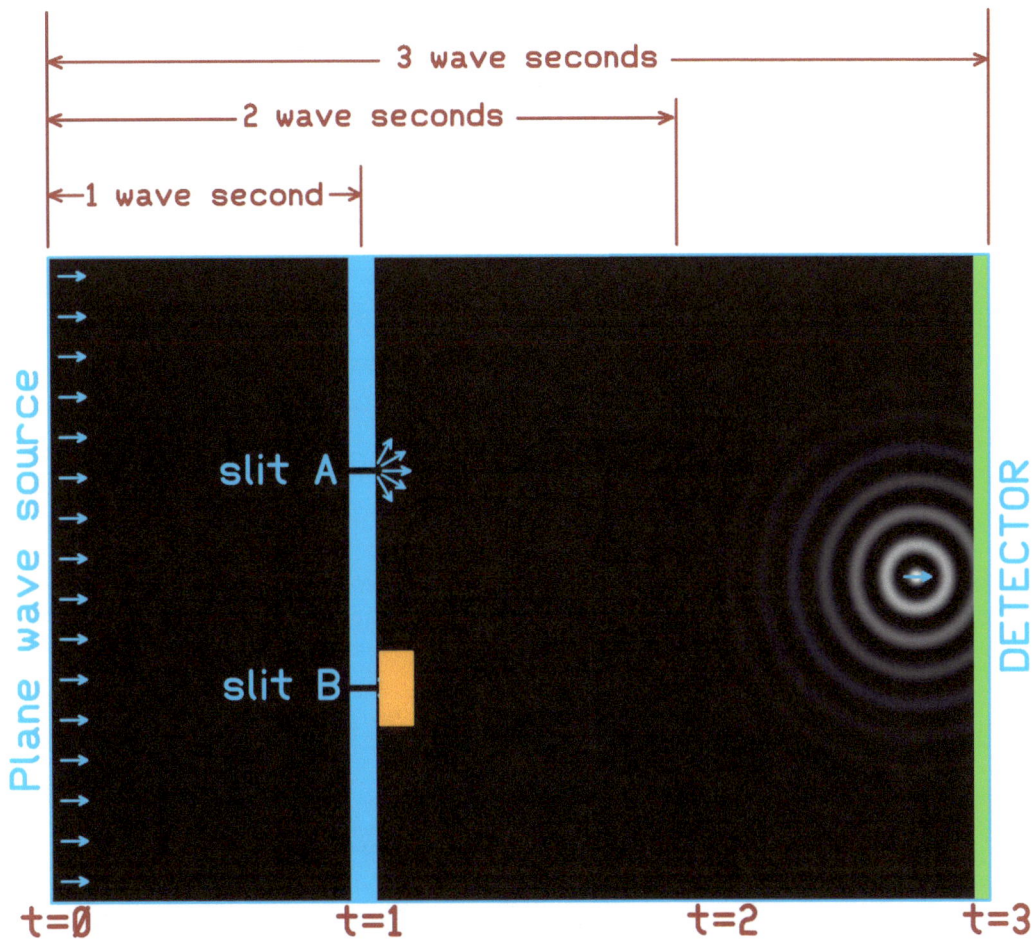

Illustration 108: Wheeler's delayed choice using electrons with the detector far from the slits.

As far as I know, no one has done this experiment. This is the exciting part of science – someone with the facilities to do this experiment will read this, perform the experiment, and publish the results no matter what they are. And as mentioned previously, scientists have the duty to do be brutally honest: if even if *I* do the experiment, I am obligated to publish it whether it supports my theory or proves it wrong.

Summary of the double slit results

I've described the main results of the series of double slit experiments that expose the heart and soul "behavior" of the universe as we know it. Quantum physics was developed to model this behavior. These words were chosen carefully – the quantum physics model can predict the results of experiments and characterize what's going on. This is the state of quantum physics today. Equations predict the behavior of the experiments but no one really understands them. Books like this one aim to change that. Today, therefore, it does not explain it in any way that goes along with our intuition. Recall what Feynman said: "Things on a very small scale behave like nothing that you have any direct experience about. They do not behave like waves, they do not behave like particles, they do not behave like clouds, or billiard balls, or weights on springs, or like anything that you have ever seen." But they do behave in a way that can be calculated and you can get experience with and form an intuition.

There are many more variations on the double slit experiment, but the experiments I've described capture most of the well known strangeness of quantum physics.

So the bottom line is that all objects, even those we normally think of as particles, have wave properties. The reason we do not notice them in everyday life is that unless the particles are very small, the wavelength is so small that the effects are too small to see. Let's go through a little of the math and see just how small the wavelength is and explore its effects for various objects.

Wave effects of everyday objects

We'll begin with red light. If you know the wavelength and the size of the slit, the diffraction angle to the first null is easily found. The sine of the angle α to the first null line can be found by dividing the wavelength of the light by the width of the slit d.

$$\sin(\alpha) = \frac{\lambda}{d}$$

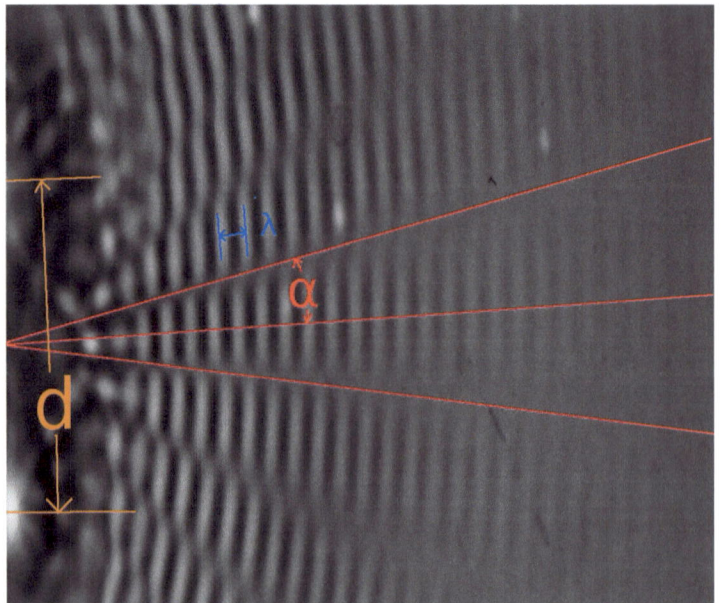

Illustration 109: Finding the diffraction angle

If you want to be able to see the diffraction, the angle has to be big enough to see. Let's suppose that with great care and patience we can see any angle greater than a tenth of one degree. $\sin\left(\frac{1}{10}°\right)$ is about $\frac{1}{600}$, so let's say that the slit can't be larger than about 600 times the wavelength. The wavelength of red light is about half a micron.   600 times half a micron is 3 mm. With great care, you can see the diffraction from red light going through a 3mm slit. If you want to see it easily, you need to be about a tenth that size, and make slits less than 0.3 millimeter wide. This can be done with a laser pointer, microscope, candle, and a razor blade in your kitchen.

It turns out that light has a definite wavelength as show in Illustration 104. What do we do for an object? They don't have a definite wavelength. But they do have a most probable one. The most probable wavelength of *any* object depends on its mass and how fast it is moving. A massive object has a smaller wavelength than a less massive object and the faster it's moving, the shorter its wavelength. The most probable wavelength λ_{object} of an object is

$$\lambda_{obj} = \frac{h}{m_{obj}\, v_{obj}}$$

where m_{obj} is the mass of the object in kg
$\qquad\quad$ v_{obj} is the velocity of the object in m/s
$\qquad\quad$ h is Planck's constant, equal to 6.626E-34 kg m²/s

Now let's see how large the slit could be for a marble to demonstrate wave characteristics. Suppose you are playing a game a marbles and would like to use diffraction to go around a marble that is in your way. See the *Wave experiments* section where we show a water wave diffracting around an object. Since a marble has a wave function, we ought to be able to do the same thing. A marble weighs about 0.04kg. During a typical game of marbles, the marble might be traveling at a speed of about .2 meters per second. Therefore, the wavelength of the marble would be

$$\lambda_{marble} = \frac{h}{m_{marble}\, v_{marble}} = \frac{6.626E\text{-}34}{(0.04)*(0.2)} = 3.3E\text{-}34\, m$$

Or 0.00000000000000000000000000000000033 meters which is very, very small. The size of a silicon atom is 2E-10 meters so your marble couldn't even diffract around a single atom in the opponent's marble! *This* is why you don't see a marble acting like a wave. But what about a really really small marble made from a single atom? If you could watch it, would it exhibit wave characteristics?

Marbles are made of glass, which is silicon dioxide. If you took one atom from the marble, it would be either silicon or oxygen. Let's check a silicon atom going at the same speed the marble went. A silicon atom has a mass of about 4.6E-26kg and

$$\lambda_{marble} = \frac{h}{m_{marble} \, v_{marble}} = \frac{6.626E\text{-}34}{(4.6E\text{-}26)*(0.2)} = 3.3E\text{-}8$$

600 times this is 4.3E-5m or about half the thickness of a piece of paper. This slit we can make, and diffraction and interference can be seen. The atom might go around a piece of dust the size of the thickness of a piece of paper.

What about a grain of salt going the same speed? A small grain of salt weighs about a milligram, so

$$\lambda_{salt \, grain} = \frac{h}{m_{salt \, grain} \, v_{salt \, grain}} = \frac{6.626E\text{-}34}{(1E\text{-}3)*(0.2)} = 3.3E\text{-}30$$

So even a grain of salt is way, way too large to diffract around even one atom.

Let's see why Isaac Newton missed the diffraction of light and ruled out light being a wave. For red light going through a 1 meter doorway,

$$\sin(\alpha) = \frac{\lambda}{d} = \frac{0.633E\text{-}6}{1} = 6.33E\text{-}5$$

$$\alpha = \arcsin(6.33E\text{-}5) \approx 0.00000633$$

The angle of diffraction is only 0.00000633 degrees and that angle is too small to see. But he did not know that $\sin(\alpha) = \frac{\lambda}{d}$, so he was not able to understand this difficulty.

What if I move really slowly?

Since λ_{marble} depends not only on the mass of the marble but also on its speed, let's turn the problem around and ask just how slowly you'd have to shoot the marble so that you could see wave characteristics and diffract around an opponent's marble.

We want the marble to be able to diffract around another marble, so we need the wavelength to be on the order of the size of a marble, about 1cm or 0.01m. Solving for velocity, we get:

$$v_{marble} = \frac{h}{m_{marble}\lambda_{marble}} = \frac{6.626\text{E-}34}{(.04)*(0.01)} = 1.6565\text{E-}030\,m/s$$

which is quite slow. Really slow. Just how slow? Let's see how long your move will take, You *have* to go at least the distance of a marble diameter, so we'll use 1cm or .01m. Using the distance formula solved for time, and making sure the units work out, we get

$$time = \frac{distance\,(m)}{speed\,(m/s)} = \frac{.01\,(m)}{1.6565\text{E-}30\,m/s} = 6.04\text{E}27s$$

Let's convert this to years:

$$6.04\text{E}27s * \frac{1\,minute}{60\,seconds} * \frac{1\,hour}{60\,minute} * \frac{1\,day}{24\,hours} * \frac{1\,year}{365.25\,days} = 1.91\text{E}20\,years$$

which is 191 billion billion years, about 14 billion times as old as the universe.

That is a long game of marbles.

The effect of large sample statistics on randomness

Toss a die in the air
to decide the cat's fate
but until it lands you don't know
so now you must wait

Illustration 110: A cat's life based on the roll of dice

The randomness in quantum physics is different from things that we normally think of as random in the macro world we live in. For example, suppose you roll a pair of dice. We normally consider that the result is random. But this is not really the case. If you know the initial position and orientation of the dice with sufficient accuracy and you know the forces that you exert on the dice with your hands, and you know the characteristics of everything that interacts with the dice such as the temperature of the air, any wind that's blowing, the bouncing characteristics of the table with the dice, etc., you can calculate the outcome. You may not have all the information you need such as the forces your hand will exert on the dice, but you could get it if you wanted it badly enough. You may not have the technology to measure all these things sufficiently accurately, but we agree that it could eventually be developed. This becomes an engineering problem of making all these measurements and then calculating the problems using classical physics. It may be a very challenging problem, but it can be done.

But this is not the case with quantum physics. It's not that we don't have enough information to calculate what's going to happen at a microscopic level, it's that you *can't* have enough information. No matter how much technology you develop, no matter how much information you gather, you cannot predict which slit a single photon will go through or when a radioactive atom will decay. No matter how much you care and how much money you throw at the problem, you cannot know the momentum and position of a microscopic particle to an arbitrary precision because the exact momentum and position don't exist. Beyond the Heisenberg uncertainty principle, there's nothing to know.

So why don't we notice all this randomness in our everyday world? Because everyday objects are large objects consisting of billions of billions of small particles. The randomness in this large sample cancels it self out in the large scale. To accept this premise, it is worth taking a look at why the world we live in does behave deterministically and why we don't notice all this randomness.

The statistics of coin flipping

Consider flipping a fair coin. By fair, I mean the coin is balanced with 50% chance of landing on heads and a 50% chance of landing on tails. If I'm flipping the coin once, there are two possibilities (H,T) with equal probability.

Now let's make some bets. If I tell you I'm going to flip the coin a bunch of times and I ask you how many heads to expect, the best answer you can give me is "about half." If I flip it a thousand times, your best guess is "about five hundred." I then press you for a definition of "about." After a bit of discussion, we agree that "about" will mean plus or minus five percent. If I flip the coin 100 times, you are willing to bet that there will be between 45 and 55 heads. If I flip it five hundred times, you are willing to bet that there will be between 225 and 275 heads.

Here is the wording for the proposed bet:

> I'm going to flip a coin a number of times. With each flip we'll write down whether it was a head or a tail. When we're done, we'll count the number of heads and divide it by the total number of flips. If that number is greater than or equal to 45% and less than or equal to 55% of the number of flips then you win the bet.

Are you making a good or a bad bet? Take a moment and write down whether you should take this bet.

_____

What did you write? Did you think it was a good bet, did you think it was a bad bet, or did you feel you didn't have enough information to know? We shall soon

123

find out that you need another important piece of information: you need to know how many times I'm going to flip the coin. To see why, let's take a look at different numbers of flips.

One toss

If you made the bet with me as is, I would choose to flip the coin once and I'd win the bet.

Why? Well, if I flip once, it will be either heads or tails. So the percentage of heads will be either 0% or 100% and never between 45% and 55%. These are the only two possible outcomes. There is a 50% chance that there will be 0% heads and a 50% it will be 100% heads. Let me put that into a fancy graph:

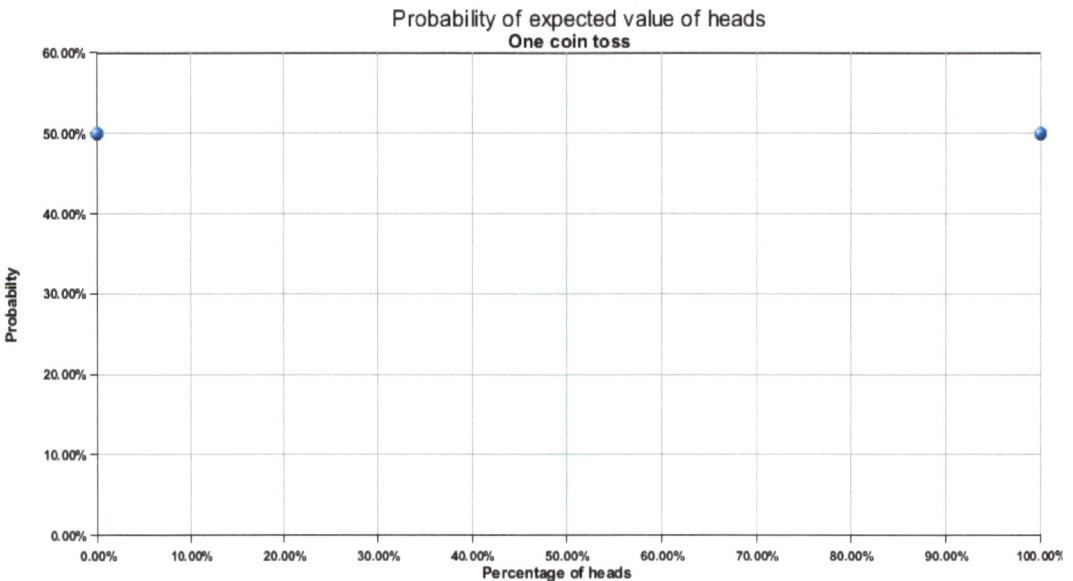

Illustration 111: Probability of getting expected value for coin one toss

The blue dot on the left side says that there's a 50% chance that it will be 0% heads. The blue dot on the right side says that there's a 50% chance that it will be 100% heads. The fact that there are no other dots says that there are no other possibilities. The shaded region shows the region where you win the bets. No shaded blue dots = bad news for you.

Two tosses

Now consider flipping the coin twice. This is equivalent to flipping two coins. Although the results are the same, it is a little easier to talk about flipping two coins than flipping a coin twice, so we will talk as if we are flipping two coins. But if we flip one coin twice, the same math applies and we get the same results. Given this math, one could say that flipping two coins is one interpretation of this graph and flipping a coin twice is another interpretation. This is not unlike the different interpretations of quantum physics such as the Copenhagen interpretation and the many worlds interpretation.

When flipping two coins, there are four possibilities. Both coins could be tails; the first coin could be tails and the second one heads; the first coin could be heads and the second one tails; both coins could be heads. Each of these possibilities has a 25% probability. Let's express this in a table:

First Coin	Second Coin
T	T
T	H
H	T
H	H

Table 2: Possible outcomes from a two coin toss

Let's consider the number of heads. If you took the bet, you're betting that the number of heads will be half the number of tosses. Two tosses, one head. This is the *expected* value. But it won't always be one head. How often will you get one head?

There are three possibilities – no heads, one head, or two heads. There is a 25% probability that there will be no heads. There is a 50% probability that there will be one head. There is a 25% probability that there will be two heads. Another way to state that is that there is a 25% probability that there will be 0% head, 50% probability that there will be 50% heads, and a 25% probability that there will be 100% heads. So you would win the bet half the time.

One way to get this information is to notice that there is a single blue dot in the blue winning region and its value is 50%.

Here's the graph version of the same information:

Probability of expected value of heads

Two coin toss

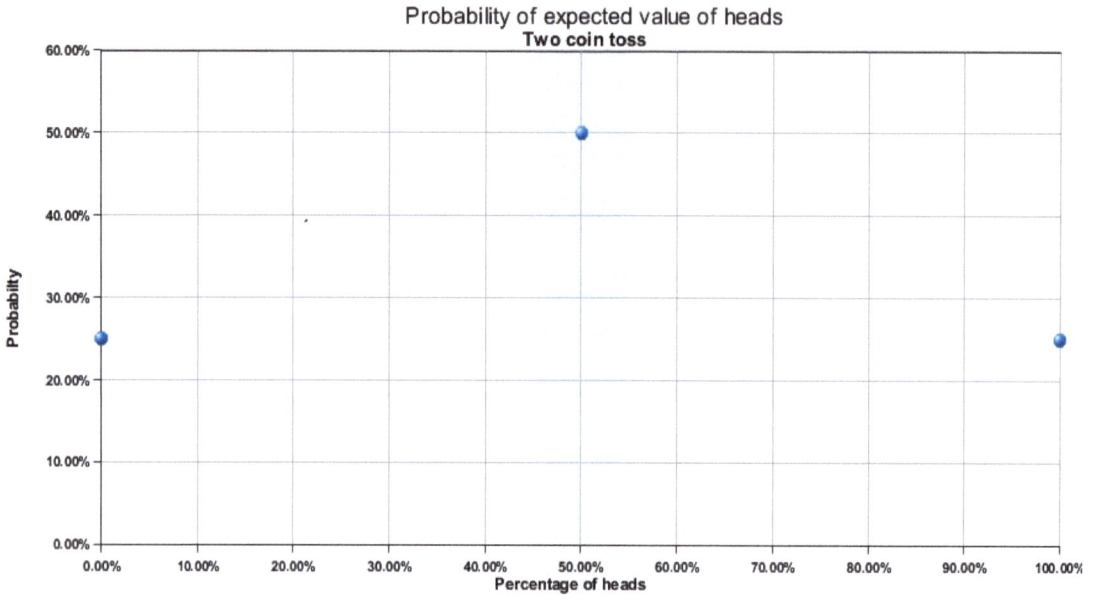

Illustration 112: Probability of getting expected value for two coin tosses

So the expected value is one head, and that expected value happens half the time.

I suppose you could call this a fair bet.

Three tosses

For three tosses, we have eight equal probability outcomes. The number of possibilities can be calculated by raising the number of choices to the number of flips. For a coin, there are two choices and we are doing three flips, so 2 raised to the 3rd power, which we write as 2^3 or 2^3, is 8. Here are the eight possibilities:

First Coin	Second Coin	Third Coin
T	T	T
T	T	H
T	H	T
T	H	H
H	T	T
H	T	H
H	H	T
H	H	H

Table 3: Possible outcomes from a three coin toss

Here is the probability of number of heads:

Number of heads	Percentage of heads	How Often	Probability
0	0.00%	1/8	12.50%
1	33.33%	3/8	37.50%
2	66.66%	3/8	37.50%
3	100.00%	1/8	12.50%

Table 4: Probability of each number of heads for three coin tosses

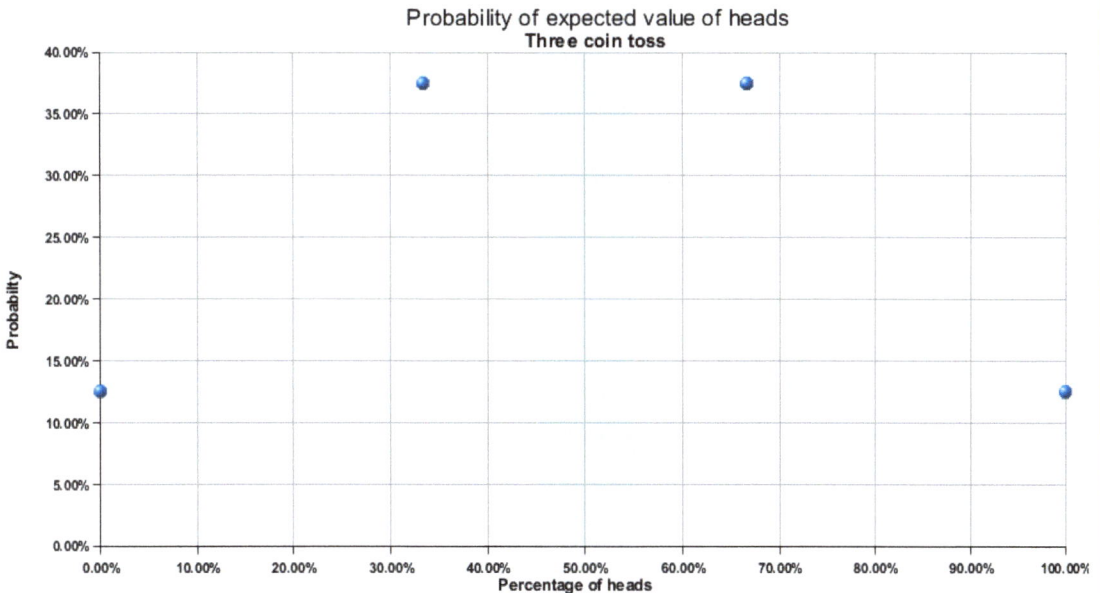

Illustration 113: Probability of getting expected value for three coin tosses

Interestingly (to me, perhaps not to you if you took the bet), the percentage of heads is *never* between 45% and 55% and I would win every time. No blue dots in the blue region.

I would not call this a fair bet.

Four tosses

For four tosses, we have sixteen equal probability outcomes since 2^4=16. Here are the sixteen possibilities:

First Coin	Second Coin	Third Coin	Fourth Coin
T	T	T	T
T	T	T	H
T	T	H	T
T	T	H	H
T	H	T	T
T	H	T	H
T	H	H	T
T	H	H	H
H	T	T	T
H	T	T	H
H	T	H	T
H	T	H	H
H	H	T	T
H	H	T	H
H	H	H	T
H	H	H	H

Table 5: Possible outcomes from a four coin toss

Percentage of heads	How Often	Probability
0.00%	1/16	6.25%
25.00%	4/16	25.00%
50.00%	6/16	37.50%
75.00%	4/16	25.00%
100.00%	1/16	6.25%

Table 6: Probability of each number of heads for four coin tosses

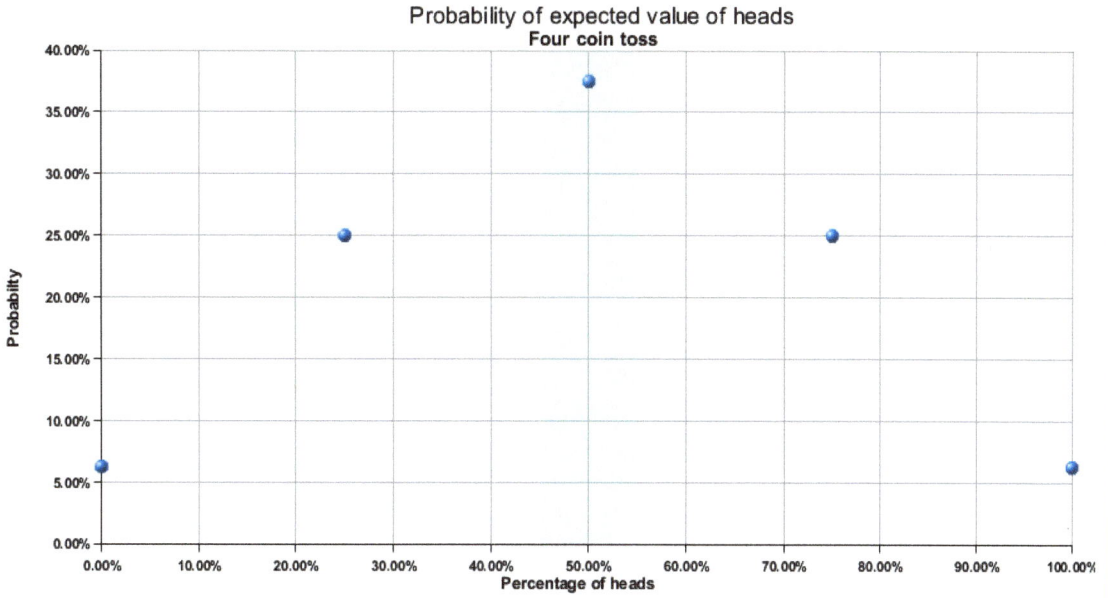

Illustration 114: Probability of getting expected value for four coin tosses

With four tosses, you'd win 37.5% of the time.

Ten tosses

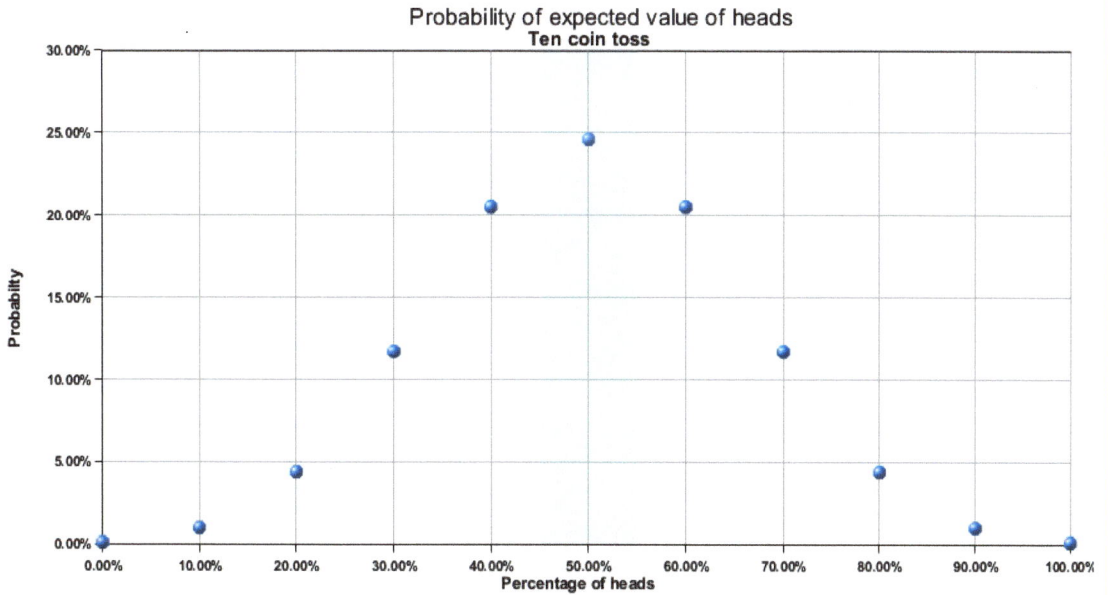

Illustration 115: Probability of getting expected value for ten coin tosses

Ten tosses would get you a win 24.61% of the time: there are 2^10=1024 possibilities. 252 of them have 5 heads and 252/1024=0.2461

Fifty tosses

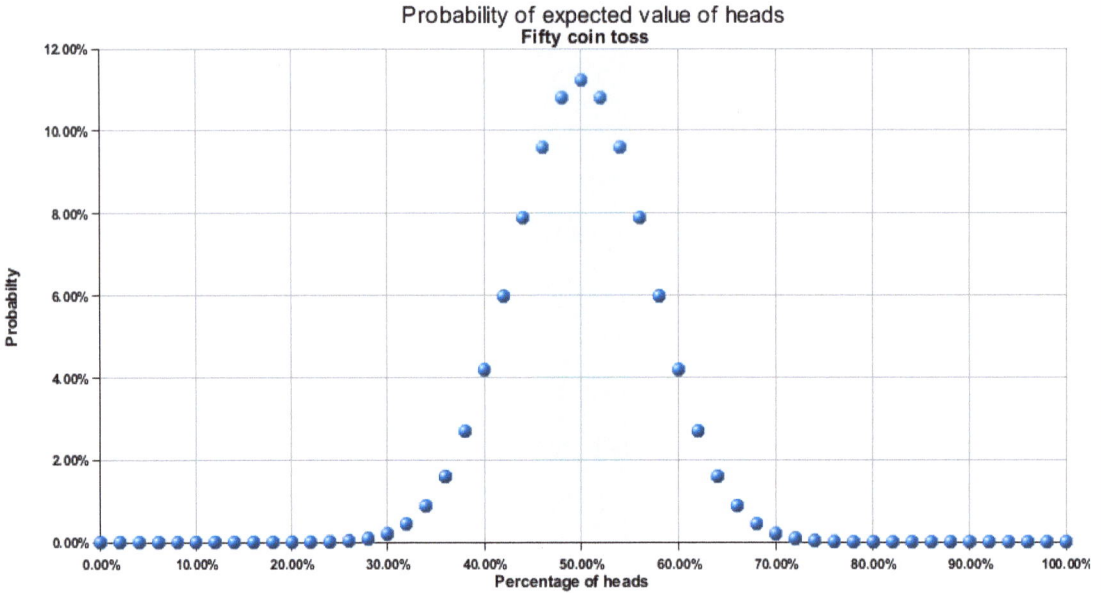

Illustration 116: Probability of getting expected value for fifty coin tosses

When the number of flips starts getting larger, you'll notice that there are more possibilities clustered around the expected value of 50%. Notice that we have five blue dots within the blue area, indicating five possible outcomes within 5% of the expected value. With fifty tosses, you would win the bet 52% of the time. This is better than anything you'll get in Vegas!

A hundred tosses

With a hundred flips, the nine dots with large probability are clustered close to the expected value and you'll win the bet 72.87 percent of the time. *Now* we're talkin'.

Notice that as we get more and more flips, the more you can count on getting close to the expected value.

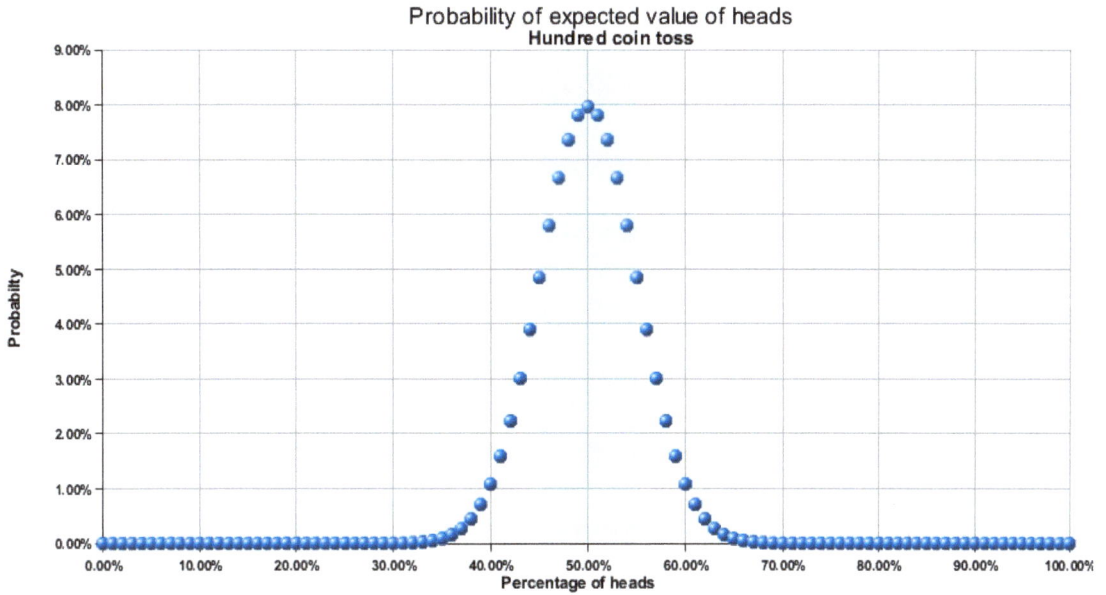

Illustration 117: Probability of getting expected value for a hundred coin tosses

A thousand tosses

With a thousand flips, all the dots with large probability are within 5% of the expected value and you'll win the bet 99.86% of the time. You'll note there are no coin flipping games in Vegas.

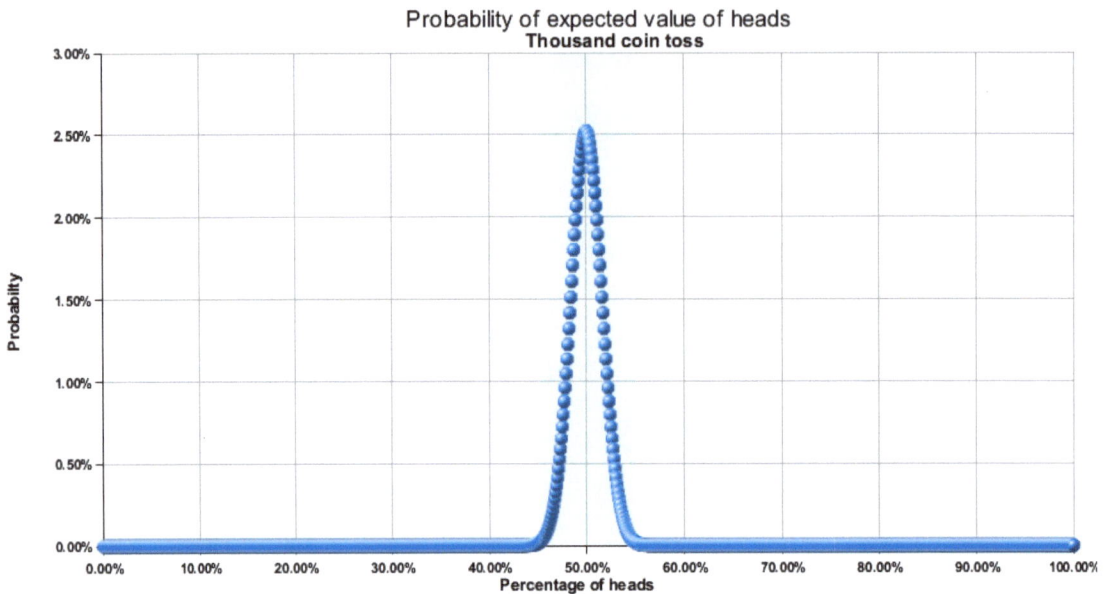

Illustration 118: Probability of getting expected value for a thousand coin tosses

Notice that as we've gotten to a large number of flips, you really can count on getting close to the expected value. These are very simple probability functions. The wave function is the square root of these probability functions. If you take a class in quantum physics you'll learn *all about* this. You are more and more likely to be equal to the expected value as the number of samples goes up.

Changing the subject for a moment, you'll read things like this:

> Suppose that you put a penny in your left pocket. If you wait one second you would expect the penny to be in your left pocket. Quantum physics teaches us that there is some randomness as to where you will find the penny and there is a chance that you find the penny in your right pocket.

Well, true, there is a chance. But it's a really really small chance. If the penny had only one atom in it, the chance isn't too bad. Given Illustration 118, if it had a thousand atoms, it's probably going to be in the expected location, your left pocket.

Taking a look at the probability functions for 1, 10, 100, and 1000 samples, you can imagine what 10,000 must look like. How about 1,000,000? How about the number of atoms in a penny? That's about 100,000,000,000,000,000,000,000 atoms. Imagine how narrow that probability function is! This tells us that it will be in the left pocket beyond any reasonable doubt.

This is why we do not experience random chaos in our everyday lives. Everything we experience is built of many many many many atoms and the statistics rule that expected outcomes will occur practically all the time.

132

The dead and alive cat

This page described the original Schrödinger's cat experiments. Because it kills a cat *(maybe)*, we took the liberty of changing the unknown aspect from whether the cat was dead or alive to whether the pet was a dog or a cat. Here was the original page.

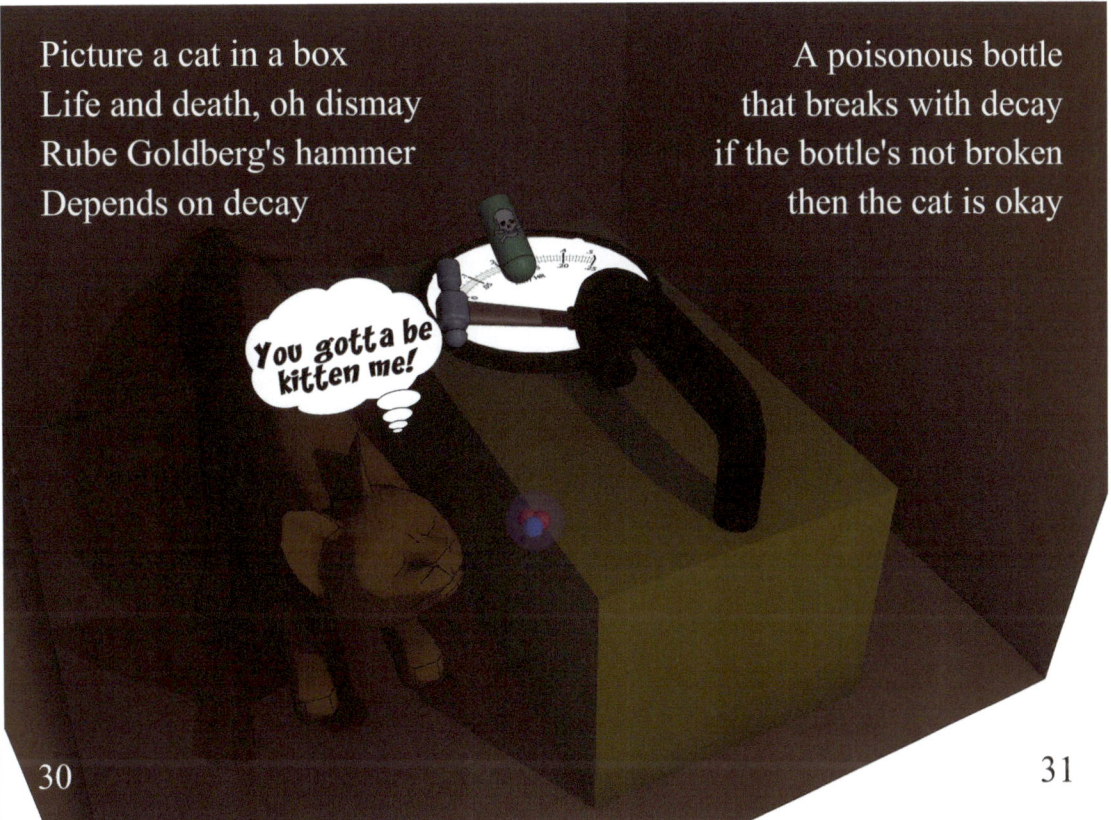

Picture a cat in a box
Life and death, oh dismay
Rube Goldberg's hammer
Depends on decay

A poisonous bottle
that breaks with decay
if the bottle's not broken
then the cat is okay

You gotta be kitten me!

30

31

Illustration 119: Original page 30-31 when we were still threatening the kitty's life

By 1935, Einstein and Schrödinger had had enough of this "crazy" quantum theory that claimed that not only were actions random, but they really didn't happen until they were observed. Much argument over this ensued and Einstein's answer to all this was his famous quote "God doesn't roll dice with the universe!"

In 1935, quantum theory of the day explained the results of the double slit experiment by saying that a radioactive atom hasn't decayed or not decayed until someone observed it. Until someone observed it, it has both decayed *and* not decayed. When it is observed, it *becomes* decayed or not decayed. It doesn't decay or not decay at the time of observation – the decay happened in the past. At the time of observation, the particle goes back and time to become decayed or not decayed. This concept is bizarre, but this theory correctly predicted the results of every experiment.

When extended into the macroscopic world such as flipping a coin, the coin is both heads *and* tails until someone looks at it, at which time it *becomes* either heads or tails. From Schrödinger's perspective, this is nonsensical and he and Einstein came up with a thought experiment that everyone could relate to.

Consider a Rube Goldberg machine that detects a particle from a decay. A particle can be detected with a Geiger counter which deflects a needle when a particle is detected. The Geiger counter has a hammer attached to the needle. The hammer breaks a glass vial containing a gas which would kill the cat. So if the atom, detector, glass vial, and Schrödinger's cat were inside a closed box, quantum theory says that until you look into the box (or smell it, or listen for purring, or x-ray it, or make an observation in *any* way), the cat is both alive and dead at the same time. When you look into the box, the cat instantaneously becomes either dead or alive but not both. But before the observation, so the theory predicts, the cat is both dead and alive.

Over the next 50 years, every experiment done agreed with the theory that the atom *has* both decayed and not decayed at the same time and that the cat would be alive and dead at the same time. Ingenious experiments such as John Wheeler's clever delayed choice experiment support the conclusion that the cat is both dead and alive at the same time.

This goes against our everyday experience at such a fundamental level that we have great difficulty accepting it. *Nevertheless...*

We don't and can't know
Not before and not now
Whether to buy another bag
of Kitty Cat Chow

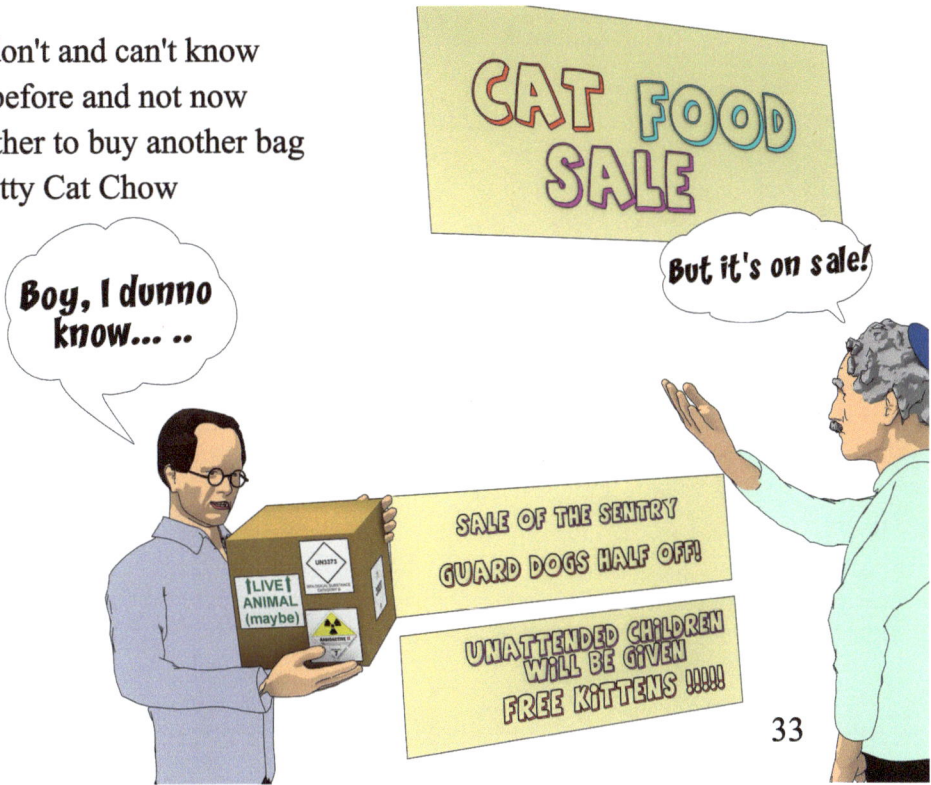

CAT FOOD SALE

But it's on sale!

Boy, I dunno know... ..

LIVE ANIMAL (maybe)

UN3373

SALE OF THE SENTRY
GUARD DOGS HALF OFF!

UNATTENDED CHILDREN
WILL BE GIVEN
FREE KITTENS !!!!!

32

33

Illustration 120: Original page of whether to buy cat food or not

This was the quadruplet when we were still killing the cat. It is funnier with the dead cat, unless of course you are explaining it to a young cat owner.

Does Relatively really affect GPS?

In the introduction, the claim was made that if relativity was not taken into account, GPS measurements would be at least hundreds of miles in error and that according to a paper from Ohio State, without taking relativity into account "the error would accumulate at a rate of about 10km per day. This kind of accumulated error is akin to measuring my location while standing on my front porch in Columbus, Ohio one day, and then making the same measurement a week later and having my GPS receiver tell me that my porch and I are currently about 5000 meters in the air somewhere over Detroit."

Well, this is true. GPS measurements are affected by relatively and if the error were left to accumulate, this would be the case. I find nothing wrong with their analysis.

Just what I was looking for

When I read Orzel's book, I was thrilled to find this Ohio State example because I was looking for an example where where knowledge of relativity allowed something that we would not otherwise have. Such an implication that you could not have GPS without correcting for relativity was perfect to back what I wanted to say in the introduction.

But both Brian Park and Richard Garwin have pointed out to me that concluding that "if we did not take relativity into account, GPS would report the wrong answer by at least hundreds of miles" is a misapplication of that analyses. This is worthy of some discussion.

Brian points out that there are many errors introduced in the measurement. Here are some: inaccuracies of the clocks; varying air density between the satellite and the GPS receiver changing the speed of light between them; clouds, rain, and any anything else affecting the propagation time; uncertainty in the position of the satellites; multipath from reflections (see the section *Challenges from multipath* in this book); roundoff errors; many more. Because of all these errors, we periodically compare the computed location of a known object on the ground to its actual known location (see the section *Experiment Controls*), find the error, and correct for the error. When we do this, we null out any accumulated error in the GPS system. It does not matter what caused the error – error from clock inaccuracies, weather, and yes relativity all get reset every time we apply this control correction.

So if we did not account for relativity, we would indeed introduce error into the system and the GPS system would not be as accurate. But in our practical system as implemented with frequent error correction, the error does not accumulate over time and we do not find reports that our porch is flying miles above Detroit.

So let's go back to my claim that we could not have a GPS system without relativity and ask a related question: if we did not know about relativity and we built GPS, would we have noticed the extra error and discovered relativity?

The answer is "probably not." Because of all the other errors, we must reset the system using the known object positions as controls so often that the error from relativity would probably go unnoticed unless someone got curious about these things and started looking at it very hard.

Richard Garwin notes that when he was working on GPS in the mid-1960s, it was before we had atomic clocks that could be put onboard a satellite, but that GPS or navigation can be done with a pulsed system (whether encoded or spread with PRN code) like GPS, but using crystal oscillators which are much less accurate and stable. How could we do this? The system had a number of receiving stations of known fixed locations that knew what the satellite should be reporting if their clocks had no errors (a control). These stations would solve for their positions, compare it to the calculated position from each of the 24 satellites, and report in real time the deviations of the individual satellite times from the individual solutions. They could then tell each satellite to correct its clock or broadcast a correction to all the GPS receivers. In fact, that is what is done to improve accuracy in Harpers enhanced GPS systems.

So, why did I leave the introduction with the egregious error? Let me digress for a moment.

Never Trust Jerry

When we teach the battery class, I'd rather students make an error and let them learn in class then than to let them go out to the field and make the error when it really matters and when it might be unsafe (some batteries are very dangerous!). For example, we do an exercise brilliantly designed by my friend Mark where we wire a battery made up of hundreds of AAA cells and power a 75 watt 12 volt bulb. These wires must carry a large current and you better use a large wire to carry that current.

Sitting on the table are some nice looking colored wires that make it easy to color code the hundreds of wires and see what you're doing. While these are tempting to use, they are too small for the application. Sitting behind these shiny colored wires are some old, dirty, black, heavy gauge wires that will do the job nicely.

In one running of the class, the students asked me which wires they should use. I simply made the comment that if you use the colored wires it will be easier to wire everything without errors because you can color code the hundreds of wires you're using. They immediately went for it and their battery failed because the wires were too small. Mark certainly knows how to bait a hook.

In the course feedback, which my boss reads, we ask the question "What is the most important thing you learned in this class?" The reported answer from student Von was, *"Never trust Jerry."*

Why is this egregious error still in the book?

This erroneous claim that GPS could not be made without taking relativity into account, is an important learning point. When you find data that substantiates what you want to prove, you must be very careful that you're relying on it because it is correct and not because it substantiates what you want to believe. It's so very easy to fool yourself when you find what you're looking for. I wrote this innocently, believing at the time that if we built a GPS system without taking relativity into account, it would report hundreds of miles of error. Finding the quote was a great feeling, substantiating what I had suspected. That gave me confidence in what I believed. But one must be so very careful not to fool oneself (see the section *Is this what we're really seeing?*).

Well intentioned people do accidentally fool themselves. Well intentioned people do accidentally fool others. And there are people who are not so well intentioned. As the Russian proverb states, "doveryai no proveryai" – "Trust but verify."

So in the interest of a lesson, I'm leaving the example in the introduction with a footnote to this section. As Von's proverb says, "Never Trust Jerry."

Actually, not just Jerry. Nullius in verba

The Royal Society's motto is *Nullius in verba* which is Latin for "on the word of no one" or "Take nobody's word for it." Don't take my word for it – reading about the Royal Society and their motto is worth some time.

Who counts as an observer?

Does the observer see first?
Or is the cat
who looks at the dice?
Is what Tegmark would ask

Illustration 121: Who counts as an observer?

The physicist asks when an outcome is determined (in physics language, they ask when the probability waveform is collapsed). The answer often given is "when it is observed."

A philosopher asks what constitutes an observer.

As did many people, Max Tegmark asked who counts as an observer. According to the Copenhagen interpretation, the cat is both dead and alive until Schrödinger opens the box and sees the cat. Max Tegmark asked, "what about the cat? Isn't the cat an observer?" Is the Geiger counter an observer because it is a macroscopic object which interacts with the microscopic object?

There is not general agreement on the answer to this question. Everyone agrees on the results of the experiments – you can't predict the outcome beyond the probabilities and more accurately than permitted by the uncertainly principle, but "what's going on" to result in the outcomes of the experiments is up to interpretation. There are many interpretations of the double slit experiment results and we will take a look at a few of them.

The situation can be summarized by describing who asks what questions. A physicist asks questions that can be answered by the scientific method (form a theory, make a measurable prediction from the theory, do the experiment and test the prediction to see if you can prove the theory wrong). A philosopher asks questions that cannot be tested.

Today we don't worry too much about who counts as an observer. Instead, we talk about when it interacts with something large enough for statistics to take over and make it appear deterministic. For more details on this, see the *The effect of large sample statistics on randomness* section.

.

Multiple Cats

For each possible choice
a universe is created
every outcome exists
is what Everette had stated

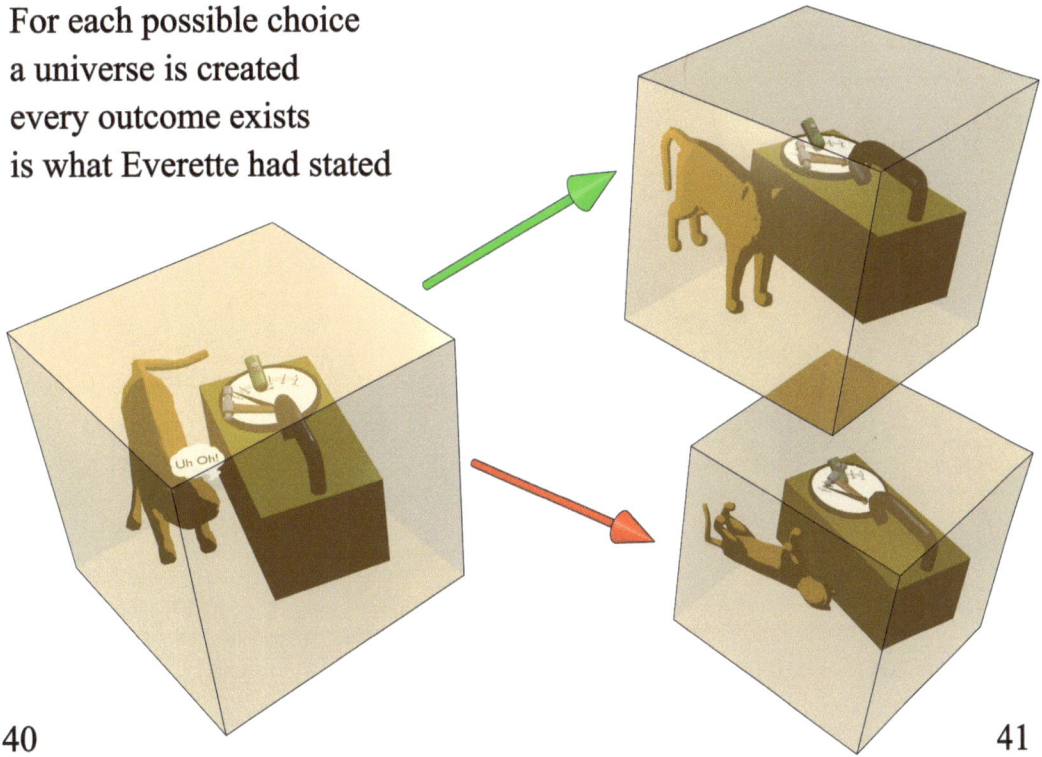

40

41

Illustration 122: Many worlds version with dead AND alive cat

Above was the many worlds page when we were still killing the cat.

The pictures on the next page were reward posters rewarding the reader with a rich set of puns. *Some reward.*

I really liked the line about Ohm not resisting on the next page. I have yet to find a second admirer, however.

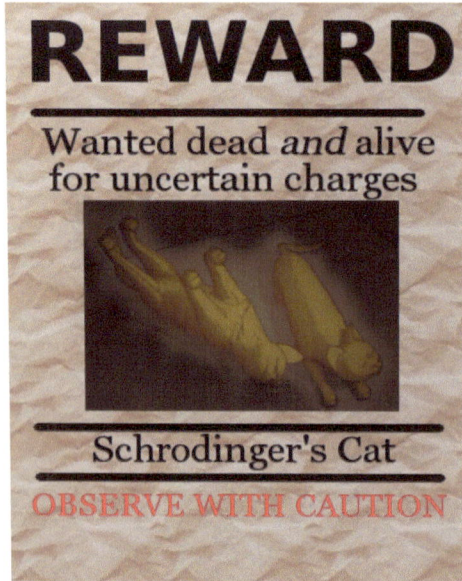

REWARD

Wanted dead *and* alive
for uncertain charges

Schrodinger's Cat

OBSERVE WITH CAUTION

But there is one more theory
for us to discuss
as strange as them all
on this crazy bus

Wanted dead *and* alive
They thought by committee
They thought up this stuff
in Copenhagen, the city

44

45

But there is one theory more
for us to discuss
as strange as them all
on this crazy bus

Wanted dead *and* alive
Copenhagen insisted
It's ironic that it was Schrodinger
Not Ohm who resisted

REWARD

Wanted dead *and* alive
for uncertain charges

Schrodinger's Cat

OBSERVE WITH CAUTION

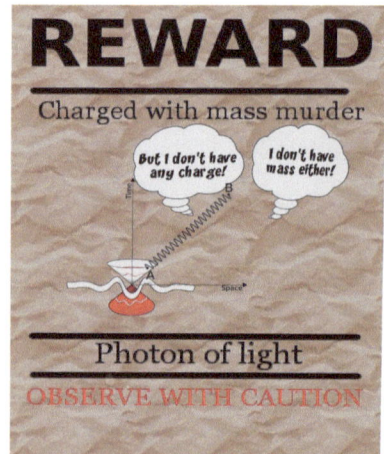

REWARD

Charged with mass murder

But I don't have any charge!

I don't have mass either!

Photon of light

OBSERVE WITH CAUTION

44

45

Illustration 124: Two reward posters

Having Fun

Illustration 125: The birth of a Nobel prize

This page was removed because it was deemed not relevant to the story. But it has a nice learning point, so I decided to include it here and end on this point.

When Feynman was doing research and teaching at Cornell, he grew tired of working so hard. This wasn't a big surprise because he had worked so very hard for four straight years at Los Alamos and had recently gone through the tragedy of losing his wife to Tuberculosis. He decided it was time to take a break from working so hard on specific research projects and just play.

He was in the cafeteria one day and saw a guy fooling around, tossing a plate in the air. As the plate went up into the air spinning around, it also wobbled. Feynman noticed the wobble and the spinning and got curious what the relationship was between the wobbling and spinning. So, just for fun, he worked out the relationship, the equations that described this dual motion.

When his friends and colleagues asked him why he was working so hard on this, he said he *wasn't* working! He was just playing around. He found this fun and kept working on it. He soon noticed that these equations would apply to electrons moving around atoms. Taking it a step further – he was still having fun, so why stop? – he worked out the equations and new techniques for small particles, such as photons and electrons. Soon he had invented a new way to compute quantum mechanics. This included his Feynman diagrams, and led to the work which earned Feynman the Nobel prize in Physics.

People who love their work are successful because they play.

Alphabetical Index

145

146

www.ingramcontent.com/pod-product-compliance
Lightning Source LLC
Chambersburg PA
CBHW041453210326
41599CB00005B/239